DIAPHYSICS

Troy Earl Camplin

University Press of America,® Inc.
Lanham · Boulder · New York · Toronto · Plymouth, UK

Copyright © 2009 by
University Press of America,® Inc.
4501 Forbes Boulevard
Suite 200
Lanham, Maryland 20706
UPA Acquisitions Department (301) 459-3366

Estover Road
Plymouth PL6 7PY
United Kingdom

All rights reserved
Printed in the United States of America
British Library Cataloging in Publication Information Available

Library of Congress Control Number: 2009926588
ISBN: 978-0-7618-4648-2 (paperback : alk. paper)
eISBN: 978-0-7618-4649-9

∞™ The paper used in this publication meets the minimum
requirements of American National Standard for Information
Sciences—Permanence of Paper for Printed Library Materials,
ANSI Z39.48-1992

Table of Contents

Preface

Math and Epistemology	1
Metaphysics	15
Physis	21
On Health and the Holy	75
Paradox	83
On the Creation of Complexity in the Universe	91
A Fractal Model for Emergence in the Universe to the Metahuman	105
Bibliography	119
Index	127
About the Author	133

Preface

This book is in part a response to the challenge set forth by Stuart Kauffman in his book *Investigations*. When we think of the laws of nature, we think of the least complex aspect of reality: physics. This is a consequence of centuries of reductionism (as are the philosophical and artistic movements and economic and social experiments of the 20th Century). But recent developments in the science of systems, complexity, chaos, bios, information (especially as it evolved into semiotics) and emergence have suggested yet another set of laws of nature that result in the creation of more and more complex things in the universe.

While the reductionist tries to understand society through people, people through biology, biology through chemistry, and chemistry through physics, what is not fully addressed is how we went from a physical universe to a chemical one, to a biological one, to a mental one, to a social one (one could properly put "to a human one" next, as it was a social species of ape which gave rise to humans). As the expanding universe has cooled, it has crystallized out into ever-more complex entities. But how? Are there rules for this kind of crystallization just as there are rules for chemical crystallization? I believe there are. They are what I am calling "diaphysical rules," the rules, or laws of nature, which result in the emergence of more complex entities from less complex entities as they self-organized.

The reductionist, mechanistic worldview had wonderful results. It gave us modern science, a great deal of material well-being (especially in places where this worldview was rejected as a political, social, and economic model, as we will see), and advanced technologies ranging from cars to computers. However, its shortcomings also became obvious when it was applied not just to simple levels of reality, like physics, chemistry, and molecular biology, but also to complex systems like governments, societies, economies, and the arts. In the 20th Century, we saw governments reduced to single-ideology dictatorships; attempts were made to eliminate churches, organizations, and even families to reduce society to single ideologies and its smallest components in the individual, who were in turn expected to dissolve into society; economies were centralized, planned, and controlled; the arts became increasingly iconoclastic until images were eliminated altogether; life was seen as meaningless, knowledge impossible, and the mind and ethics as illusions. The end result was the slaughter of hundreds of millions of people by this dehumanizing worldview. All we are left with now is the hopeless nihilism of postmodernism. That is the end result of the reductionist, mechanistic worldview.

All of this could be reason enough to reject reductionism—or at least question it as a complete description of the universe. It is the latter which has happened with the development of emergentist science. This science had its origins with people like Ludwig von Bertalanffy, whose ideas were developing prior to World War II, but which were published only after the war (one's ideas, as Bertalanffy knew, most be *timely*), the information theorists, and the development of game theory. In the 1960's and 1970's, chaos and catastrophe theory contributed. Bios is an even more recent contributor. The scattered pieces of an emerging paradigm have slowly self-organized (in a way predicted by the emerging paradigm itself) into works like Stuart Kauffman's *Reinventing the Sacred* (whose preface could almost stand in for this one), Frederick Turner's *Natural Religion*, and many of the works listed in this book's "Bibliography and Suggested Readings." So why the need for *Diaphysics*?

There are several elements I believe I am contributing to the new paradigm. One is the consideration of the wave-nature of the universe and its enfolding as fundamental to complexity. This unites questions of energy, matter, and time. I also bring together (and into the paradigm) J.T. Fraser's emergentist view of nature and time experience wit the human psychosocial emergentist theory of Claire Graves, Don Beck, and Christopher Cowan, explaining them in light of each other and the new emergentist paradigm. I am also, as the title suggests, laying out an argument for a set of natural laws that manifest themselves in different ways at different levels of complexity, which then give rise to new levels of complexity. These are laws which run *through* the different levels of reality. Thus the term *diaphyiscs*.

Diaphysics is fundamentally a work of philosophy, and is aimed primarily at those in the humanities. This may seem strange to say, since it is full of math and science. Well, one of the consequences of reductionism was the separation of the sciences from the humanities; one of the consequences of emergentism is their reunification. Science was once known as natural philosophy, and that's what this work is. Since this is a work of philosophy, and philosophy is supposed to be contemplative in nature, I have arranged this work in aphorisms (some quite lengthy) to encourage contemplation between each one. My hope overall is that this work will prove beneficial to those in every field who deal with systems and complexity, and not just those in the humanities. And while I do hope that in the end I am right about what I have written in this book, I at the very least hope that if I'm wrong, that I'm wrong in some very interesting ways.

Naturally, a work like this was developed not in isolation, but in dialog with other. The first person I must thank, then, is Stuart Kauffman, whose work *Investigations* was the seed of *Diaphysics*. I must certainly thank my friend and colleague Kristen Hauck, without whom many of the ideas in this book could have neither been born nor developed. I must also thank Frederick Turner and Alexander Argyros for their encouragement and for the positive contributions they have made to the way I think (all negative elements of the way I think are personal idiosyncrasies, and should not be blamed on anyone but me, my own DNA and my own habits). The members of the International Society for the Study of Time (ISST), particularly the ideas of J.T. Fraser have also contributed a great deal to this work, some directly, some indirectly, through conversations and articles I've read. I would also like to thank Lou Marinoff for his helpful feedback. Finally, I have to thank my wife, Anna, for all the support and encouragement she has given me.

Math and Epistemology

1.

Radical skepticism says we cannot know anything. The fact that we have survived as a species belies this claim, since those who could in fact not know anything could not know about the lion creeping toward them. This person was eaten. So obviously we can know some things about the world—we can recognize predators and prey, as well as trees to escape into, cliffs to avoid, and weapons to throw and dodge. But these are objections to radical skeptics only. We can only be open to knowledge if we are both skeptical and willing to accept reasonable evidence

2.

What is reasonable evidence? The postmodernists are now already accusing me of logocentrism. Yet even postmodernists give reasons why they believe what they do. And they typically try to provide some sort of evidence. Their actions belie their claims, which are only poor excuses for sloppy scholarship. Also, they get to defend indefensible positions by creating hermeneutic circles that no one can get into or out of. Well, as they say, don't argue with a drunk person, so I'm not going to spend too much more time on the postmodernist rhetoricians.

3.

Reasonable evidence to a mathematician consists of creating a proof. Reasonable evidence to a physicist or chemist consists of creating data from a reproducible experiment. Reasonable evidence to a biologist consists of data from reproducible experiments and statistical models. Reasonable evidence to a social scientist consists of statistical models and expert interpretation. Reasonable evidence to someone in the humanities consists of textual analysis and expert interpretation. Reasonable evidence to a religious person consists of the truth as either revealed to him through mystical experience or as told to him by a holy man and/or book. Reasonable evidence to an average person consists of evidence of the senses and testimony of experts. All of these are reasonable evidence, depending on what the question is. Reasonable skepticism keeps us asking questions.

4.

I am going use a few philosophical terms here that won't trouble philosophers much, but which need definition for non-philosophers. I use the Greek terms *physis* and *nomos* because they include more than the English terms "nature" and "the mental." *Physis* includes all the physical world from quantum physics through chemistry and including the body and the brain. *Nomos* includes all the products of the brain, including mind, culture, society, economy, and the arts and humanities. More specifically, it is the product of several interacting brains. *Nomos* emerges from the interactions of the human brain level of *physis*. These human interactions—I removed "brain" because brains only work when embodied—are possible through *logos*. *Logos*, another Greek word, can be translated many ways, but all of them boil down to the concept of information communication. We will return to these ideas again.

5.

Social constructionists suggest we can make the world into anything we want. If this is true, then we can fundamentally alter *physis* as we wish. By all means, refuse to believe in the oncoming car. But the oncoming car believes in you and will, nonetheless, run you over. Just because we do in fact create certain elements of our reality—the value of money, for example—it does not then follow that we create all elements of our reality. In its weakest forms, social constructionism has some validity. We create the value of money, but the value of money exists in the realm of *nomos*, which is the realm in which we create reality. But if we think that paper money will continue to have value once we burn it, then we are ignoring *physis*, to our own detriment. The example is a lesser form of what tragedy teaches us about the relationship between *physis* and *nomos*. And when the proper art forms are not available to teach us this lesson, we unfortunately have to learn it the hard way, as millions of Russians learned when the Soviet Union accepted the ideas of Lysenko, and tried to grow wheat in the tundra. It is to help prevent the enactment of such tragedies in the future that I am engaging in this investigation into the nature of *physis*.

6.

Does knowledge of something mean we can name it? Wittgenstein said in the *Tractatus* that "The limits of my language are the limits of my world." But is this true? What does it mean to say we know the sky is blue? Is it more accurate to say "I call the color of the sky "blue" in English"? Is etymology of any help here, or does it just remind us how words are arbitrary in the choice of sounds to represent actions, things, and qualities? Etymologies can show us relationships among words, provide us with metaphors with which to think, but it is limited in helping us to know. Another problem with equating language and knowledge is the fact that there are languages lacking a word for the color we refer to in English with the word "blue". Some cultures have not needed it. But if you ask someone who exclusively speaks such a language to separate colors into, say, ten different categories, they will have a group we would recognize as blue. So it seems clear that they know of blue, even if they literally cannot say "blue," lacking the word for it. Thus, the perception is there, but not the word. The limits of their language are not the limits of their world—they can perceive blue, even if their culture has not needed a word for that color. And we must not forget that words can sometimes mask the fact that we have many dissimilar things grouped together under one term—which could in fact get in the way of our recategorizing things and, thus, reconceiving them. But that is precisely one of the things artists and poets are for.

7.

Another error in epistemology is Kant's idea of the "thing in itself" which, according to Kant, we can never know anything about. One could perhaps argue that the idea of the "thing in itself" has spurred us on to further investigations into the nature of *physis*, particularly in the realm of subatomic physics, and perhaps it has. But I do believe that it, like Descartes' separation of man into body and soul in order to create a space for science to work free from the interference of the Church, has outlived its usefulness. And I hope, too, to show that there is another approach that will be just as fruitful, if not more so, for spurring continued investigation into the nature of *physis*. This does not mean, of course, that there is no such thing as the "thing in itself," whatever that may be, but since it is in a realm of what is by definition, unknowable, I reiterate Nietzsche's four theses:

> *First proposition.* The grounds upon which 'this' world has been designated as apparent establish rather its reality—*another* kind of reality is absolutely undemonstrable.
> *Second proposition.* The characteristics which have been assigned to the 'real being' of things are the characteristics of non-being, of *nothingness*—the 'real world' has been constructed out of the contradiction to the actual world: an apparent world indeed, in so far as it is no more than a *moral-optical* illusion.
> *Third Proposition.* To talk about 'another' world than this is quite pointless, provided that an instinct for slandering, disparaging and accusing life is not strong within us: in the latter case we *revenge* ourselves on life by means of the phantasmagoria of 'another', a 'better' life.
> *Fourth proposition.* To divide the world into a 'real' and an 'apparent' world, whether in the manner of Christianity or in the manner of Kant (which is, after all, that of a *cunning* Christian—) is only a suggestion of *décadence*—a symptom of *declining* life. . . . That the artist places a higher value on appearance than on reality constitutes no objection to this proposition. For 'appearance' here signifies reality *once more*, only selected, strengthened, corrected. . . . The tragic artist is *not* a pessimist—it is precisely he who *affirms* all that is questionable and terrible in existence, he is *Dionysian* . . . (Nietzsche, *Twilight of the Idols*, pg. 49)

As Nietzsche points out, it is quite pointless to even discuss what by definition cannot be known. If the "thing in itself" can only exist in a "real" world we cannot ever know anything about, then how can we even know it exists to talk about it—and why bother talking about it at all? I think Nietzsche here stakes out the most eminently reasonable position possible.

8.

Nietzsche also points out in his *Fourth proposition* what part of the problem is, though he does so in a backwards kind of way. He observes that artists only represent the 'apparent' world, but do so while selecting, strengthening, correcting what is represented. In other words, what they produce on canvas, on stage, etc. is a concept, a map, a game-board of the object. Many of our errors in thinking arise through misunderstanding the relationship between perception and conception. This situation is most egregious in the most abstract way of thinking we have developed: math. When people want to defend a transcendental worldview (particularly against science), they often bring up mathematics. It has become the final holdout for transcendental thinkers.

9.

When transcendentalists try to defend the transcendental world, the 'real' or noumenal world, they ask: "How can you scientifically prove that two plus two equals four?" Since you cannot, they claim, mathematics must be transcendentally true. However, this shows a misunderstanding of how one can use science to understand aspects of philosophy. Using the scientific method is not the only way of using scientific knowledge to prove something is factual ("true" in Nietzsche's "uninteresting" sense—from "Truth and Lies in a Nonmoral Sense"). If we can show that it is natural for the brain to process the world in such a way that we can make statements such as 2+2=4, then we have a science-based (though not "proven" scientifically per se) explanation for math, meaning we do not need a transcendental explanation for it.

10.

Math is the way we abstractly express relationships in nature. Words are sounds we use to represent conceptual categories, which are derived by observing many objects and placing objects with similarities into categories. Take the Bactrian camel. We categorize camels as either Bactrians or Dromedaries, because Bactrians have two humps, and dromedaries have one. All the Bactrian camels more closely resemble each other than any one resembles a dromedary, or any other animal, for that matter. So we categorize two-humped camels as Bactrian. But are Bactrians in fact identical? No, each one is different—we erase the differences so we do not have to create a different category for each individual object in the world, which would be very cumbersome (this, despite the fact that we do oftentimes give an individual its own category, as when we name our pets, because we become so familiar with them that they become more individuated to us). Our brains, to be more efficient, conceptualize. If brains did not do that, the owner of such a brain would not recognize that the cat that ate a member of the group was similar enough to the approaching cat that it would be prudent to try to escape. We would be like Funes in Borges' story "Funes the Memorious." This is why vervet monkeys have different calls meaning 1) big cat, 2) eagle, and 3) snake, which each results in different responses. Also, anything that eats must be able to recognize what is not-food and what is food. To eat, one cannot have to relearn this information each time. Life was much simpler for one-celled predators: eat anything that moves (or, more accurately, whatever binds to the outside of the cell, meaning there is a kind of conceptualization even at the chemical level in the surface proteins). Those who could not make these judgements about the world and create concepts would have died either from eating something poisonous or from being eaten. Any animal that could not make proper judgments regarding the reality of the world they were in

Math and Epistemology

would not have been able to survive. Those who were better able to make those judgments would be able to survive better. I give as evidence a human population of over 6 billion at present.

11.

Due to the complexity of our brains and our use of language, humans are able to create more conceptual categories than any other animal. Further, those categories can overlap, and they can exist in nested hierarchies. The Bactrian camel is simultaneously a camel, in the camel family (which includes the humpless llama and its relatives in South America), an herbivore, a mammal, a vertebrate, an animal, and alive. Thus a Bactrian camel shares similarities with other herbivores—elephants, rabbits, buffalo, manatees, etc.—in that they all eat only vegetation. We call it a mammal because it has hair, is warm-blooded, and feeds its young milk, just like platypuses, whales, koalas, and leopards. It is a vertebrate because it has a backbone and an internalized skeleton, like fish, birds, reptiles, and amphibians. And it is alive because . . . well, that's a question we still need to answer.

12.

What is a number? What is the number 1 or 2? In other words, what does the symbols "1" and "2" mean? Gottlob Frege in *The Foundations of Arithmetic* points out that ""the number one is a thing" is not a definition, because it has the definite article is on one side and the indefinite on the other" (I). One could ask in what way is "1" a "thing" (solving his problem of definition)? What definition of "thing" could it meet? I think we can at the very least agree that numbers are nouns, seeing as they meet all the criteria of being nouns. Numbers can be made plural and possessive, we can use a derivational ending *(-ness)*, and they can function as adjectivals or adverbials. Now, if we use the way we teach young children what a noun is, as a person, place, thing, or idea, then we can see what classifications numbers could have. Certainly they are not persons or places—which leaves us with things or ideas. Let us next agree with Frege that a number is not a thing. That leaves "idea." So a number is an idea, not a "thing." More, it is a universal. But making this statement only leads us into what W. V. O. Quine in *From a Logical Point of View* calls the "ontological problem of universals: the question whether there are such entities as attributes, relations, classes, numbers, functions" (9). The problem with the way Quine has stated the problem is that none of these are "entities." There is no "entity" "1". Number is an attribute of a thing or a set of things, and attributes are informative. Attributes, features, relations, classes, numbers and functions all provide information about entities. Quine presupposes a materialist ontology, and these are certainly not material. Worse, they are the attributes of adjectives, adverbs, and determiners. When we use numbers in sentences, we use them as determiners: "I see two camels." This is an attribute of the set of camels, their "twoness," and not of the camels themselves, as we would have in: "I see wooly camels." "Wooly" is a feature of the camels' topology. "Two" is a feature of the set. So perhaps we should start again.

13.

"Two" is a conceptual category. It is necessary to keep track of group members (we are a social species after all), and to make proper divisions (as chimpanzees do when sharing meat from a kill). It would also be useful information if one is hunting or gathering. "Yes, there are two of them. We need more hunters." "Two" is a conceptual category in the same way as "Bactrian camel" is. If we have two humps on two different camels on two mountains with two rivers flowing from each mountain, then the commonality among humps, camels, mountains, and rivers is "twoness"—a conceptual category we designate by using the word "two". The sound "two" may be an arbitrarily chosen sound to represent this concept, but the concept, and the fact that a word exists to represent it, are not arbitrary. Quine points out that abstraction adds more things to the universe—the multiple entities still exist, but now the one term used to talk about them all as one is added (70). Now, if abstraction does add something to the universe, what it adds is information. The multiple stages of a river constitute something inform, which we put into a form by refering to it as a particular river—the Caÿster, to use Quine's example. It informs the universe, adding form to what was inform. Thus, the universe is complexified. Names are abstractions for a single entity that changes in space and time (such as the Caÿster River). A concept is an abstraction for many entities that are similar, but separated by space and time (ex: river, red, square). And, "once abstract entities are admitted, our conceptual mechanism goes on and generates an unending hierarchy of further abstractions as a matter of course" (Quine, 78). "Number" exists as a high level of abstraction.

14.

Numbers express quantity. This statement belies Frege's that "with the numbers we have difficulty in finding even a single common property which has not actually to be first proven common" (15). The concept of quantity is not even uniquely human. Parrots, chimpanzees, and dolphins all understand the concept of quantity. Humans have, with language, coined words which name particular quantities. Over time we in the West have simplified our numbers, from Roman numerals to Arabic numerals which, with the zero, allow us to create a pattern of numbers which repeat. We could base such repetitions on any base: our clocks use base 12, our computers use base 2, and in everyday arithmetic, we use base 10. Once we have done this, we can more easily see patterns. But these patterns are in part of our own making. With base ten, we have the following logical formula: we start with zero then, using the formula $x_{next}=x+1$, we use iterative transformations (which are very common in the universe, creating fractals) to get to 9. Then, in the next iteration, we bring back the zero, moving it over with the first number to create the next set of ten. Thus, beginning with zero, we count down 0-9, and across 0-9 to get, across, 0, 10, 20, 30, 40, 50, 60, 70, 80, 90. Having reached nine again, we simply repeat the process of adding the next digit. Thus we loop around to 100, and lead into 110, 120, 130, etc. across, and 101, 102, 103, etc. down. In an experiment with 4 year olds in a pre-K class, the children learned how to count to 100 very rapidly using this method of learning numbers—suggesting this is the most natural way for us to derive the next number in a series (also known as counting). Thus, the number chart derived from this method, which starts with zero at the top left hand corner, counting down to 9, and across from 0 to 90, has proven very useful in teaching children as young as four to count and learn their numbers very rapidly—more rapidly than when one uses the standard chart that goes 1-10, then repeats 11-20, etc., which masks the pattern. Also, despite Frege's complaint about J. S. Mill's notion of addition as piles, these same 4-year-olds have learned how to add by counting piles of objects and putting the piles together. Doing this, they have learned to transfer this skill to counting and adding together other objects, showing 1) they understand the concepts, and 2) that perception of objects precedes concept-development.

15.

Most of the mistakes Frege makes in thinking of number and the origin of number comes from the fact that he mistakenly thinks primitive people faced the same problems and lived in the same world as Frege himself did. The concept of and words for number originated millennia before Frege, and so occurred in a world where "The number word "one," again, in the expression "one straw" signally fails to do justice to the way in which the straw is made up of cells or molecules" (30) is irrelevant. It is likely number came about the way color words came about (here again, we will see where Frege is wrong regarding color and number as being utterly different). Just as there are languages which have only "one" and "many," there are languages which only have "black" and "white" as color words. In color words, we see a pattern of addition of color words: first is red, then yellow or green, then green or yellow, etc. as the language users need more color words. Equally, as the need to enumerate increases, more number words are added. How does one express the fact that there are a lot of a food animal, a few but enough to put together a hunting party, or not enough to bother with? We have many, some, not enough. Or how do you get enough in a hunting party? What if you need "one more," but you don't really want to choose because you don't want to make someone angry? The need to enumerate occurs because we are a social species, and we live in ever-more complex societies. The more complex the society, the more complex the numerical system. We see enumeration coming on full-force with trade, when people needed to know what was lent or borrowed, bought or sold. It seems likely more complex enumeration came about at about the same time as writing, as both are needed to deal with an ever-more complex society. Historically, analogue estimation systems precede larger sets of exact numbers. As Steven Pinker observes in *The Stuff of Thought,*

> Many cognitive scientists believe that the human mind inherited two systems for keeping track of quantities from our mammalian ancestors. One is an analogue estimation system, in which quantities are gauged in an approximate manner by relating some continuous magnitude in the head, such as a vague sense of "amount of stuff," of the extent of an imaginary line. The second system keeps track of exact quantities, but only up to a small limit, around three

or four. Neither of these is adequate to thinking about quantities that are both exact and large, like 9 or 37 or 186,272. For that, one needs to learn a number system in childhood and arithmetic operations in school. (129-30)

In fact, "More sophisticated systems capable of tallying exact large numbers emerge later, both in history and in child development. They tend to be invented when a society develops agriculture, generates large quantities of indistinguishable objects, and needs to keep track of their exact magnitudes, particularly when they are traded or taxed" (138-9).

16.

Let us return to Frege's "one straw." Frege says that there is never any benefit to considering something to "be incapable of dissection" (43) and that doing so will always result in a false conclusion. This position is naive to the extreme. It neither considers the fact that primitives may in fact not see a difference (not knowing that the straw is made up of cells), nor does it take into consideration that Frege himself does not consider people mere collections of cells, but as unified into individual people, any more than he takes a rock to be a mere collection of atoms only, but a whole rock—that these are *necessary* lies we tell so we can speak about anything. "One straw"? Of course! The straw is not merely a collection of cells or atoms, but something with emergent order—the atoms and cells add up to something.

17.

Frege: "number is neither spatial and physical . . . nor yet subjective like ideas, but non-sensible and objective. Now objectivity cannot, of course, be based on any sense-impression, which as an affectation of our mind is entirely subjective, but only, so far as I can see, on the reason" (38). In this statement Frege makes two mistakes: he separates the objective from the subjective, and he separates the senses from reason. Both are fallacious. One cannot separate the subjective from the objective—one can only be more or less subjective and/or objective—and one cannot separate the senses from reason. One cannot reason without sensory input, and all things that sense have at least some degree of reasoning (all living things have reasons for what they do; they do not act in random fashion, nor are they completely determined automata). Further, when Frege objects that a concept is not "something subjective like an idea" (60), he does not recognize that so-called objective concepts arise due to commonality of experience and commonality of language within a language group—that we teach our children to have common words for common experiences. When we all agree on common experience, that is what we call the objective world or an objective concept.

18.

John Stuart Mill is correct to note that we derive the concept of 3 empirically, from our naming a collection of objects * * * as being 3 in number. He is further correct in noting that we derive arithmetic from the combination of * *, or 2 objects, and *, or one object, to form a collection of 3 objects. What Mill does not state directly, and which Frege himself does not grant, is that from the observation of many collections of different kinds of objects, we develop the concept of 3. Thus, even though Frege claims that using Mill's empirical derivation of number, "we can see that it is really incorrect to speak of three strokes when the clock strikes three, or to call sweet, sour, and bitter three sensations of taste; and equally unwarrantable is the expression "three methods of solving an equation"" (9-10). Certainly the first two examples are not visual —but they are empirical in that they are detected using the senses. More, though, they constitute 3 in Mill's sense because of what Mill leaves out, and which he assumes the reader to understand (there is a lesson to be learned here regarding assumptions, even for philosophers), which is that naming different collections of objects similar enough to be considered to be in the same collection (conceptually similar) of the same number with the same name is the creation and naming of an object.

19.

Simple addition based on concepts derived from empirical information allows us to then logically continue counting up to numbers which would include numbers that do not represent actual collections of objects, but would include imagined collections of conceptual "objects." This process of concept formation allows us to conceive of things we cannot perceive, such are our powers of abstraction. Anthropological research shows some tribes to have number names only up to two or three. Yet, they are able to conceive of larger numbers when asked to do so. They do so because the rules for deriving small numbers—numbers we may consider very small—can give us larger numbers, whether we have either the names for them or the objects to represent them. For us, 10,000 is a large number, but for tribes with few number names, 10 is such a large number they do not even have a name for it. That does not mean there are not collections of objects for either one that approach that number (consider the collection we designated with Avagadro's number, 6.023×10^{23}—the number of molecules in a mole), or that we need actual collections to derive large numbers.

20.

No zero? Frege suggests that if we accept Mill's number empiricism, "the number 0 would be a puzzle; for up to now no one, I take it, has ever seen or touched 0 pebbles" (11). True, but there have been plenty of people who have been in plenty of places where pebbles are not present. We have a language which designates absence. "Is Bob in the house?" "No, Bob is not in the house." Frege implies we cannot really say this, since we cannot see or touch no Bob in the house. Frege would have to eliminate all but affirmative propositions from logic, and it would be impossible to say "No dogs are cats." The fact is, we can observe that something is not there, precisely because, and only because, we have empirical information. We have seen and touched such objects in the past, and we can thus notice when we cannot see or touch such objects. Incidentally, Western mathematics had to make do without 0 until Arabic numbers were introduced. This did not mean people did not have a concept of no things— there was just no mathematical term or number word for it. Arithmetic with

Roman numerals is difficult, but not impossible. Simplicity is one of the benefits of Arabic numbers—as well as their creating more obvious and shorter patterns than did Roman numerals—but we must not mistake these patterns for being transcendental just because the patterns are simple and obvious and depend on 0 as a starting point for the pattern.

21.

An equation expresses relationships. Relationships are inherent in nature. Math describes nature so well because all of nature is relationships—each object has and is in a relationship with other objects (objects here include pure, substance-less energy). "Equals" is such a relationship. If there are a number of objects that we call (in English) "two", and we have another number of similar objects that we would also call "two" due to the number of objects being equivalent, then we have a number of objects that we designate in our language as "four". If "four" represents this many objects: * * * * , and two represents this many objects: * * , and 2=2, then 2+2=4. A transcendental explanation is not needed. All we need is a proper understanding of how concepts are formed in the brain, and we can learn that through cognitive psychology, which is science. Thus, while science cannot prove using the scientific method that 2+2=4, while 2+2 cannot equal 5, and never can as long as we use the language and notation as we presently do, I have shown that a proper understanding of science can help us use philosophy to understand the source of mathematical statements as non-transcendental.

22.

It is not "numbers" which can be divided, but members of a set, the quantity of which is represented by numbers. Thus, sets with even numbers of objects can be divided evenly: 4x can be divided into 2 sets of 2x objects, or * * * * can be made * * and * *, where * * = * *. Even animals understand the ideas of equal, more, and less. Thus, those numbers which represent a quantity that can be equally divided are called even. Those which cannot are called odd. We can then recognize that there is a pattern in these distinctions on the number line, which allows us to develop the concept of even and odd numbers up and down the line. The same is true of other properties. We can place these objects: * * * * into a square: * *, with each side containing * *. Thus we get that 4 = 2 square. We
* *
can also see that we get 3-squares and 4-squares: * * * and * * * * , respectively.
* * * * * * *
* * * *

We count the number of objects, see that they are 9 and 16, respectively, meaning a 3-square = 9 and a 4-square = 16. The concept of square too is developed by observing many objects of such a shape, which was called a "square," the properties of which were then worked out. Like all concepts, the square is an idealization created in the mind out of many objects in the world with similar features whose differences were erased. This, too, is how memory works, or we would live the cursed life of Borge's Funes.

23.

Frege objects to Mill's understanding of the symbol + by saying "that if we pour 2 unit volumes of liquid into 5 unit volumes of liquid we shall have 7 unit volumes of liquid, is not the meaning of the proposition 5+2=7, but an application, which only holds good provided that no alteration of the volume occurs as a result, say, of some chemical reaction" (13). Frege fails to recognize that there are in fact two kinds of arithmetic: addition of equals and of unequals. Frege is making the mistake of thinking that

$$5x + 2x = 7x$$

which is what Mill is actually referring to when he discusses the empirical origins of arithmetic, is the same thing as

$$5x + 2y = 7z.$$

Mill is talking about adding 5 of the same things to 2 of the same things to get 7 of the same things. With a chemical reaction, we add together two different things.

$$NaOH + HCl \longrightarrow NaCl + H_2O.$$

Frege would have us believe this is the same thing as adding together 1 liter of water and 1 liter of water to get 2 liters of water. A balanced chemical equation is not the same as the adding together of identical things. If we pour 2 ml. of water into 5 ml of water, we will get 7 ml of water each and every time. It is a property of the universe.

24.

Mathematics is how we abstract out relationships we see in nature—it is the ultimate expression of the similarity of unlike things. One objection that may be made is that such facts as this one, that the sequence of odd numbers adds up to the sequence of whole numbers squared: first odd number = 1; 1 squared = 1; second odd number = 3; 1+3=4; two squared = 4; third odd number = 5; 1+3+5=9; three squared = 9 shows that numbers are somehow transcendent. However, this is still not transcendental. It is an expression of relationships that actually exist in the physical universe. We can just recognize, then abstract out, those relationships. The error is in thinking that our concepts are prior to our perceptions. When we think our concepts are prior to our perceptions, we think there must be a transcendental world. However, it is perception which gives rise to conception—thus, transcendentalism is not necessary as an explanation. In transcendentalism, the one gives rise to the many, while I am suggesting that the many give rise to the one. I find that much more beautiful than any transcendental "explanation," aside from it being based on what we have learned about how the brain works. I hope too to show through this work how a mathematics based on observation of macrophysical objects is equally applicable to explaining such things as the world of quantum physics. This would explain the 'good fortune' we have that mathematics can and does explain the world very well, even though we must always remember that math only gives an exact approximation to an inexact (approximate) world—therefore, one must use careful

25.

Frege objects that W. S. Jevons "would rewrite the equation 3-2=1 in some such way as this: (1' + 1'' + 1''') - (1'' + 1''') = 1'. But what would be the remainder of (1' + 1'' + 1''') - (1'''' + 1''''')?" (49). Shall I go so far as to say that here Frege really shows he does not understand arithmetic at all, or at the very least that it necessarily intersects with the real world (something that should be clear enough by now)? The 2 must be included in the set 3 for 2 to be subtracted from it. This is where transcendentalist philosophy leads us: into considering the subtraction of 2 things other than those in the original set from the original set. This idea is corrected using set theory. A set is a species containing individuals of that species. Thus, when you subtract 2 from 3, it must be two within the set of 3, not external to it. This is implied in the formulation of 3-2=1 as 3x-2x=1x, where we subtract out the "x" to get the conceptual 3-2=1. The set theory of arithmetic allows us to understand numbers and arithmetic conceptually—and Jevon's idea also makes clear the unity in variety and variety in unity in math (making it a beautiful theory, when we define beauty with Francis Hutcheson as unity in variety and variety in unity). Frege objects that Jevon's view that "distinct ones and twos and threes," with his superscript notation "is utterly incompatible with the existence of arithmetic" (57-8). However, this is true if and only if arithmetic has nothing to do with the physical world—there is always the presumption in doing arithmetic that we are talking about one or two or three of something or of different objects. They can be unexpressed, but we are always talking about objects in a set.

26.

We are enamored with the idea of a transcendental world because such a world is an unchanging world. This is why we reject the evidence of the senses, even though it should be obvious that if our senses lied in any significant way, we would not have survived as a species to even be discussing such matters. This is why I side with Nietzsche when he says:

> I set apart with high reverence the name of *Heraclitus*. When the rest of the philosopher crowd rejected the evidence of the senses because these showed plurality and change, he rejected their evidence because they showed things as if they possessed duration and unity. Heraclitus too was unjust to the senses, which lie neither in the way the Eleatics believe nor as he believed—they do not lie at all. It is what we *make* of their evidence that first introduces a lie into it, for example the lie of unity, the lie of materiality, of substance, of duration. . . . 'Reason' is the cause of our falsification of the evidence of the senses. In so far as the senses show becoming, passing away, change, they do not lie. . . . But Heraclitus will always be right in this, that being is an empty fiction. The 'apparent' world is the only one: the 'real' world has only been *lyingly added* . . . (Nietzsche, *Twilight of the Idols*, pg. 46)

In a proper presentation of the nature of *physis*, therefore, we must discuss at length the issue of time. The passage of time is not an illusion of the senses, as

Math and Epistemology

many physicists, including Einstein, would have us believe. The issue is much more complex, as we will see.

27.

Only if conception precedes perception can it be true that addition "does not in general correspond to any physical relationship," as Frege claims; nor is it true that "It follows that the general laws of addition cannot, for their part, be laws of nature" (14). Even if it is a mental process, that mental process is natural.

28.

Why worry about arithmetic? Hector Sabelli makes it clear in *Bios*:
> It is cogent to start with the simple. Arithmetic functions are the simplest. Also, fundamental aspects of nature are quantities. Action is a multiple of Planck's quantum, and information is contained in the difference between successive actions. Information is thus portrayed by subtraction and addition. The number series itself represents quantity and order, contains the basic algebraic structures (lattice, group, topology), and involves a logi that cannot be reduced to something simpler than arithmetic (Gödel's theorem). Numbers also represent basic forms. (84)

So as we can see, if we are to talk about anything regarding the nature of the world, we have to start at the beginning, with arithmetic. And that means, too, that we have to have a proper understanding of the nature of number and arithmetic. What kind of edifice could we construct with a poor foundation? The foolish man builds his house upon the sand.

29.

Hector Sabelli observes that the simple gives rise to the complex (that is, to ordered complexity, which is the kind of complexity we are concerned with here, not disordered complexity, which simply amounts to randomness). So this, again, is why we start with mathematics. There is nothing simpler; which may seem an odd thing to say since so many seem to have difficulty with math. First, many people think that math is concrete and things like love are abstract. In fact, as we have partially seen, the opposite is true. But love is more complex, and we have been taught, especially in the West, to prefer the simple over the complex. But things that are truly more complex have more concrete reality than do the simple.

30.

Math is simpler than physics, which is simpler than chemistry, which is simpler than biology. Physics deals with fewer objects, fewer interactions, fewer relationships, and fewer rules than does biology—which is why it is simpler. Humans evolved in a complex world of social interactions. When faced with simplicity, we try to find complexity in it, and only complicate things. It is the disconnect between the expected and what is there that creates confusion and distress when people try to do mathematics. Not everyone is comfortable thinking so simply.

31.

There are plenty of good reasons why we have chosen to believe in a 'real', noumenal, unchanging world of the Forms, or Ideas, even though such is almost certainly not the case. Or, at least, it is utterly impossible to prove either way. Since it is impossible to prove a negative (that it does not exist), it is up to the transcendentalists to prove it does exist. Since such a world has been defined as one that, by definition, we cannot know anything about, I chose not to waste my time discussing it. I do believe, however, that I have shown that what little evidence that has been brought to bear in favor of a transcendental world to date has been shown to be nothing more than placing the conceptual cart before the perceptual horse.

Metaphysics

32.

What is metaphysics? Most Western philosophers have translated the "meta" of metaphysics as "over" or "beyond." The metaphysical is over or beyond the physical—it is what gives form to the physical by being over, beyond, and outside of the physical. This way of looking at the metaphysical is why such philosophers as Heidegger and his existentialist and postmodernist followers have rejected metaphysics (though Derrida also points out that we can never actually get outside of metaphysical thinking). But there is another way of thinking about metaphysics where the metaphysical is not outside of physics, but is, to use another translation of "meta"—the one Aristotle meant in calling his book after *Physics*, *Metaphysics*—as "after." If we think of the metaphysical as that aspect of reality that comes "after physics," then I think we can recover this child from the expelled bath water.

33.

What do I mean that "metaphysics" comes "after physics"? We can begin to see what I mean from John Casti's comments in his book *Complexification* regarding the objections mathematicians and physicists make about using ideas from math and physics to explain issues of society (that is, using *physis* to understand and explain *nomos*, an issue to which I will return shortly): "In philosophy, the term *metaphysics* refers to the study of questions that transcend the merely natural, issues such as whether every event has a cause and what things are genuinely real" (77). He uses catastrophe theory, a mathematical theory, to explain the collapse of civilizations. But to do so,

> One begins by singling out a handful of variables, calling them inputs. These quantities, in turn are *postulated* to give rise to observable changes in another handful of quantities that we term outputs. And to keep things within the confines of elementary catastrophe theory, we further *posit* that there are no more than six inputs and two outputs. Next we *assume* that whatever the mathematical relationship linking the inputs to the *outputs*, the analytic structure of that relationship satisfies the technical conditions required by the Classification Theorem. In broad terms, these conditions amount to the requirement that the system's outputs be fixed-point attractors of some smooth dynamical system. Finally, we *presuppose* that the coordinate systems used to measure both the inputs and the outputs happen to coincide with the coordinate systems that lead to the standard catastrophe geometry governing a system with the number of inputs and outputs we have chosen. (79-80)

The details are not relevant to this particular discussion. What is relevant is what Casti says in his next paragraph:

> Now focus your attention on the italicized words in the preceding paragraph. Each one of these words represents an out-of-the-blue assumption that we must be ready to swallow if we want to appeal to the Classification Theorem to single out a

particular geometry describing our problem. It's at this point that physics gives rise to metaphysics. (80)

In the move from physics to metaphysics, there is a tacit assumption that the models by which we understand the physical world are relevant to understanding issues of culture, government, art, and literature—the realm of *nomos*. And here is where metaphysics becomes truly "after physics," as we come to see the world as a nested hierarchy of different levels of complexity, starting, loosely, with physics and, moving up through the biological, then into the realm of the human mind and human invention.

34.

By the time of Aristotle, Western philosophy became divided into *physis* and *metaphysis*. In the time of Heraclitus, the division was between *physis* and *nomos*, which is why Nietzsche set out to reformulate the issue as Heraclitus had, after millennia of post-Aristotlean metaphysics. In the time of the ancient Greeks, *physis* was more or less nature (to use the Latin term), which included human nature. *Nomos*, from which we get "nomenclature," means "naming," "law," "culture," and includes the arts and literature. If metaphysics is formulated as "after physics," we could say *nomos* and metaphysics are the same. We could, but I think that would be a mistake. It would be better to say that *nomos* and metaphysics should be the same, but only if metaphysics is understood as arising out of physics, and not the other way around.

35.

New metaphors are needed. So let me reformulate these ancient philosophical terms in modern, scientific terms. Let us agree that *physis* includes what is now modern-day physics, chemistry, and biology. One could then use these terms in the following way: physics, for this discussion, includes physics and chemistry (we do this because those objects considered by mechanistic macrophysics are objects made of molecules—a ball rolling down a plane is a ball and a plane, both made of molecules—or are large groupings of atoms, as in stellar plasmas); (meta)physics for biology (I do this because some physicists have difficulty applying physical theories to biology—this is not surprising since biological objects are a new, emergent reality and, thus, use their own rules; but even these rules do abide by certain physical rules, as we shall see); and metaphysics for anything that emerges due to human thought (including governments, economies, laws, ethics, art, and literature). Biology, including neurobiology, is the bridge between physics and the products of human minds. The other reason I say biology is (meta)physics is because biology emerges from physics and is literally after physics. I put "meta" in parentheses because we must still consider biology in the realm of *physis*, even if we are, in a sense, updating these terms.

36.

What, precisely, should we call theories such as catastrophe theory, information theory, chaos theory, emergent properties theory, etc. that can be used to understand physical, (meta)physical, and metaphysical models? Casti says one uses these theories metaphysically to deal with certain problems, but if it is legitimate

Metaphysics

to use these theories to explain things at these different levels of reality, there must be a sense in which we could call these theories themselves metaphysical. The first answer to that is that to do so is to reintroduce the idea of metaphysics as beyond physics, as an overarching rule. But this is precisely what I want to avoid. These rules themselves originate in the realm of physics, and get passed on to each new level, evolving and expanding to form that new level. A strange attractor in simple fluid dynamics is similar, yet different, from a biological strange attractor. One set of biological strange attractors, for example, creates each type of cell in a multicellular organism. Another set of strange attractors makes that collection of cells a rhinoceros, a sheep, or a human being. Water dripping from a faucet at a certain rate does so in a pattern created by a pair of strange attractors; a human being is shaped into a human pattern by at least 130 strange attractors. This shows that it is not just the number of attractors, but their complexity as well. Water drips in a chaotic pattern; humans are biotic patterns.

37.

We need a new term for these kinds of laws that go through *physis*. Perhaps a good term would be "diaphysics," from the Greek *dia-* for "through," found in "dialog" and "dialogics." A diaphysical rule would be a rule that is propagated through the different levels of both *physis* and *nomos*, through the physical, the (meta)physical, and the metaphysical. This avoids the problems of seeing the rule as somehow above physics, outside the physical world, yet helps us understand that it is found at every level of reality. It also ties the idea to dialogics, bringing us to the idea to *logos*. We can think of contemporary information theory as *logos* realized in contemporary terms. Insofar as we are coming to see that everything in the universe is made of information, we can connect it to Heraclitus' philosophy and his statement that "It is wise, listening not to me but to the *logos*, to agree [*homologein*] that all things are one" (Kahn XXXVI). If everything is information (*logos*), then everything is, in fact, one—even if everything in the universe is realized (informed) into many particulars through the informing action of information. And yet, information is also inform—it is insubstantial—thus, it is nothing as well as one, two (digital and analog), and many. This is how strange attractors, the absent centers which create and form the systems through the actions of the elements of the system itself which emerges, are realized. So information is physical, diaphysical, and dialogical. It is *physis*, giving rise to *nomos*, both of which are forms of *logos*, or information.

38.

If we look to evolutionary biology, it makes sense to position *logos* between *physis* and *nomos* when it comes to humans, as truly human culture, including art and literature, only developed after the evolution of language. Biology gave rise to language, which gave rise to the richness of human culture. There are certainly some aspects of *nomos*—certain cultural elements—that precede human language, but even so, such cultural elements could not have evolved without some sort of intelligent communication. Information was passed on, even if it was not passed on as rapidly and efficiently as it has been since language evolved. That is why chimpanzees and bonobos have far fewer, and far less developed, cultural elements than do humans. Pre-human cultural elements, once developed, do not change much over time. Change is an important aspect of human cultures. In fact, as new information technologies were developed, we have seen those cultures which adopted those new technologies evolving much faster than they did before the development of those technologies, and much faster than did those cultures that did not adopt the technologies. Thus, the boom in European culture after the invention of the printing press. And with the advent of telecommunications, especially of the Internet, we have and will continue to see even more rapid cultural change, giving us more cultural choices than we have ever had. With the creation of more choices, we will have increasingly more freedom—freedom is choice, and the more choices we have, the more freedom we have (one must be careful here, as one could argue that we should have the freedom to choose dictatorship—but that is a false choice, as it is one that prevents, and even reduces, future choices). We must be careful, though, that we do not mistake political anarchy for freedom. Anarchy/disorder and micromanagement/rigid order are opposites with the same results—freedom is neither disorder nor order, both disorder and order, simultaneously.

39.

If we are to understand *physis*, *logos*, and *nomos*, we must first understand what the diaphysical laws are. Using them, we can better understand *physis*. But we must not think the diaphysical is the same thing as *physis*. We will do well to look at both the diaphysical laws and at the specifics of our evolutionary past as apes, primates, mammals, vertebrates, animals, and living organisms. By looking at ourselves through the diaphysical and through *physis*, especially our (meta)physical aspects, we can better understand both *logos* and *nomos*.

40.

It is important we do not make the mistake of applying all aspects of *physis* to the human—the mistake many have made in trying to apply Relativity Theory to ethics. An explanation of gravitation as bending space, of space as curved, and of the relativity of the passage of time depending upon the speed at which one is traveling does not logically lead to thinking of ethics as relative—meaning, in the most radical versions of this world view, that there is no such thing as ethics. While we are indeed in space and time, Einstein's theory explains elements of the cosmos that are at the extremes—in this case, extremely large and extremely fast objects. At the same time, while we are indeed ultimately made of quantum particle-waves, which constitute atoms, quantum physics is an explanation of the other extreme of reality—the extremely small. In size, humans exist between these two extremes, while at the same time we are at the pinnacle of physical complexity (so far), so we should expect to have other kinds of rules for humans. In the same way, mechanistic physics describes very simple, middle-sized systems (mechanistic physics also break down at the extremely large and the extremely small and the extremely complex—the complex being systems with more than two physical elements to the system). We should not expect

Metaphysics

(though the entire philosophy of naturalism was based on this deterministic view) humans, which are at the opposite extreme level of complexity to mechanistic physics, to follow the rules of determinism, either. We should, however, expect the diaphysical rules that order systems at all these levels to apply equally to humans, including our language and our culture, economies, political systems, and art and literature, as these will be the rules that exist at each level, no matter how complex the level may be (though it should be noted that it is diaphysical processes which drive the creation of complexity in the universe).

41.

Tragic art teach us that if *nomos* does not map onto *physis*, if *nomos* does not become metaphysical, as Casti defines it, we can expect tragic consequences. When you reject this version of metaphysics, you have decided you want man to be tragically destroyed—or at least severely punished for having listened to your ideas. The great works of tragedy, including those of Aeschylus, Sophocles, and Shakespeare, show us what happens when our *nomos* does not map well onto our *physis*—that is, when our *nomos* ceases to be metaphysical in the sense I have outlined above. The benefit of works of tragedy is that they allow for a smooth transition from one set of ethics and worldview to another—from an archaic worldview to a median one. But when we do not have works of tragedy to warn us of the dangers of extending *nomos* beyond the bounds of *physis*, of abandoning metaphysics, we end up enacting tragedy in the real world—as we did in the 20th century with the Nazis (national socialism) and the Communists (international socialism). Both were anti-metaphysical *nomoi* whose supporters attempted to deny the existence of *physis* altogether, or to adjust *physis* to their worldviews (as we saw with Lysenko's biological theories, which were anti-Darwinian and anti-Mendelian, since both were seen as bourgeois theories). It is ironic that the Nazis and Communists both embraced technology, as the Nazis were anti-technology (they embraced technology to try to win the war, but they had an essentially agrarian worldview), and both they and the Communists were anti-science—or, at least, anti-biological science (biology refutes and refuted Communism—and Nazism, for that matter). In fact, much of the anti-science attitude among many of our humanities scholars today comes directly out of Heidegger's pro-Nazi, anti-science, anti-technology views. This is also the origin of our theory of political correctness (which claims that if we change our language, we can change our nature, since we in fact have no nature, according to the post-Heideggerians), which is in fact a top-down theory and, as such, a metaphysical theory in the old sense, where *nomos*-language can re-create our *physis* (the gods may have fled for Heidegger, but the ability to speak all of reality into existence has not—it has just been given to man).

42.

One element of our *nomos*, technology, does allow us to manipulate the world in such a way that we can create new things—but this is only *physis* informed by *nomos*. There are elements of reality that can be and are affected by language, but each of these—marriage, money, contracts, etc.—are well within the realm of *nomos* and are thus part of the reality we do create. The mistake is in thinking that *nomos* and *physis* are in fact the same thing, meaning if we can change *nomos* by speaking it, then we should be able to change *physis*, including that part of *physis* known as human nature, by speaking it as well. Language is nowhere near that powerful. The evolution of language certainly had an effect on the evolution of humans immediately after language emerged, but we will note that the effect was only on the most immediate level of *physis*, the level closest to language. It does not affect our deepest biological drives, our biological needs, our chemistry, or our physics. Perhaps future technology will change that, and we will, through *nomos*, be able to affect our *physis*. But will what we produce even be human?

43.

We need to better understand our past and get our present straightened out before we can move on to a healthy future. We need to understand the physical, the (meta)physical, the metaphysical, the diaphysical, and how to make our *nomos* better match each of these, if we are to become healthy, if we are to avoid enacting tragedy in the real world.

Physis

44.
Information is *logos*; it is the *archae* of all things. Through emergence *logos* became flesh.

45.
Everything in the world is information. This statement is in complete agreement with Heraclitus' idea that everything is *logos*, and *logos* is all and one. If everything is information, that means that all matter and energy is information (though lack of energy, for example, can also provide information, so the inverse is not necessarily true). In other words, I am proposing an ontology of information. If we accept an ontology of information, many of the statements made by quantum physics are clarified. However, the second law of thermodynamics says that entropy increases over time. If information is the opposite of entropy, information in the universe is being lost. And if matter and energy are information, matter and energy in the universe is being lost. This violates the first law of thermodynamics, which states that energy cannot be created or destroyed, but only transformed. So one of these laws appears to have to be incorrect, as most currently understand both laws. Either that, or the universe is not a closed system. We too often forget that the laws of thermodynamics as commonly understood are only applicable to a closed system.

46.
Is there anything more fundamental than information? Sabelli suggests action precedes information. But we also know that action = energy x time, so both energy and time must precede action. Further

> Power spectrum analysis relates energy with time. Power is the statistical frequency of action quanta; temporal frequency is 1 over time. As a rule, physical energy is a function of frequency: the higher the frequency, the greater the energy.
> In 1/f patterns, the greater the frequency, the lesser the energy (power). (*Bios*, 151)

This relation between action, energy, and time results in power law spectra, which are the organizing principle of self-organization. Self-organization is thus a fundamental feature of the universe.

47.
Jeffrey Satinover observes that

> Out of the rule-constrained but absolute randomness of [a] quantum system there seems to be the potential for an unlimited number of bits of information. Absolute chance therefore implies infinite information. Each time a wholly uncaused quantum event comes into existence, it's a new piece of information in the universe that was not there previously; it wasn't in the least implied by a chain of prior events. (141)

Therefore, there is an ever-increasing amount of information in the universe. And information is the opposite of entropy. So why do things seem to be running down and burning out universe-wide? If both information and entropy are increasing in the universe, should they not be balanced out, somehow? Or do things only seem to be increasing in entropy because of our highly informed perspective?

48.

Action bifurcates into information, which interacts with and informs action to create matter. Action, information, and matter are fundamental to all levels of organization and complexity in the universe.

49.

"Information is physical" (Charles Seife, *Decoding the Universe*, 2). Let me repeat that: information is physical. Information "is a concrete property of matter and energy that is quantifiable and measurable" (2). Information is measured in bits — quantum information is measured in qubits. "It from bit." Or, more accurately, "It from qubit." All systems require information to exist, since information communication allows for co-operation.

50.

"Thermodynamics is just a special case of information theory" (Seife, 72). Relativity and quantum mechanics "are actually theories of information" (119). Living organisms are information-processing machines, and work to maintain and reproduce the information in their DNA. But brains, especially human brains, "are constantly acquiring and adapting to information that they have gathered from the environment. The human brain is an information acquisition machine as well as an information-processing machine" (114). It is information all the way down.

51.

When Nietzsche found an apparent contradiction between the first and second laws of thermodynamics, he developed the Will to Power to reconcile the contradiction (if reconciling contradictions does not sound like Nietzsche, keep in mind that we are talking about a true contradiction, not a paradox or complementary opposites, which I will discuss and, like Nietzsche, affirm). That is one approach to solving the problem—one which, if we truly understand what Nietzsche means by Will to Power, is a reasonable approach to solving this problem.

52.

To say information and entropy are opposites is not to say that we necessarily lose information, but that our ignorance of the universe increases over time. The information of the universe is increasing, but it is doing so at such a rate that we can only become increasingly ignorant of the state of the universe, due to the information increase. If entropy is merely a measure of our ignorance, we have returned to the issue of epistemology. This means that the actual information content of the universe is increasing, and is increasing at a rate faster than we are able to know what has been created. This is one way of solving the problem of entropy increasing in a world clearly increasing in complexity. Sabelli observes that

> The standard model postulates decay towards disorder, but new analyses of statistical entropy presented here [in *Bios*] demonstrate that entropy measures symmetry and diversity in the data. Entropy does not measure disorder or decay. Processes spontaneously generate complex patterns and structures rather than equilibrium and uniformity. In the course of cosmological evolution, energy becomes matter. Abiotic evolution naturally continues into biological evolution rather than tending to entropic disorder. (*Bios*, 9)

We have a mistaken view of entropy because "Entropy does not measure disorder or decay," but rather, "Entropy maximization implies greater diversity and symmetry" (459), as bios theory shows.

53.

Another argument is that the second law, as commonly (mis)understood, only applies to closed systems, such as engines. Open systems, where more energy is always entering, by definition do not run down. The earth's ecosystem is such an open system—the sun provides the extra energy. Dissipative structures are open structures (the ecosystem as a whole is a dissipative structure), and living organisms are dissipative structures. Order is maintained because energy comes in, is transformed, and released. Now, in the case of dissipative structures, the second law still applies—the energy released has less information content than what came in. The earth is getting its energy from the sun, which is entropic in that it is burning up all of its fuel. Thus, the sun is a closed system, feeding energy to the open earth ecosystem. Dissipative structures are local decreases in entropy at the expense of increases in entropy elsewhere. More, Sabelli argues that in biotic systems, "Entropy decreases with the generation of bios" (459). However, it is not clear whether the universe itself is a closed or open system (though Sabelli's evidence that the galaxies are biotically distributed does seem to argue for an open system). If it is expanding (and it certainly seems to be doing just that), then it is likely an open system. Particularly if the universe itself contains the principle of growth. But if the universe is mechanistic (I make this argument knowing full well that it is not), then the expanding universe is like a container full of air with a stopper—the universe expands, everything in the universe merely spreads out like gases in an expanding container, and entropy thus increases. It seems there are many who fundamentally see the universe in this way. But, as I said, this is the mechanistic view—which we have learned is not applicable for either the very, very large or the very, very small. And these two extreme ranges are what are most important to understanding the cosmos. Mechanistic physics is useful for good-enough approximations of middle-range sized physical objects. Newton's calculations are good enough to get the space shuttle into orbit—here relativistic equations would give a level of accuracy unnecessary for the project at hand. But mechanistic physics is a special case, and a good-enough approximation that causes us to develop inaccurate models of the universe—or complex systems—when adopted as a worldview. We still think mechanistically even though we know this kind of physics is only useful in special cases—notwithstanding that we have done a lot of very interesting things with this special case.

54.

Then there is the argument Frederick Turner makes in his work *The Garden* that since we are highly complex, everything else in the universe looks, in relation to ourselves, entropic. Since we contain the most information, other entities at hierarchical levels below us all look entropic to us. Here we have another epistemological problem—that of perspective. Again, if we understand that we are taking this perspective, we can correct for it, to see what real relationship exists between information and entropy.

55.

A speculative question: if something more complex than humans came along, would it appear to be negentropic? Would we be able to "see" it at all?

56.

Finally, there is the argument from bios. Hector Sabelli says information is created "in the form of new patterns and structures. Expansion, innovation, and diversity are characteristic of physical evolution. Diffusion and diversification represent irreversibility. In contrast, mechanical processes are irreversible; they conserve information and maintain their pattern," while, when a process converges to an attractor, "information is lost; once the process hovers within the attractor, information is conserved" (152). But this is only the beginning. Biotic processes, which are characterized by bipolar feedback and the creation of novelty, turn out to be one of the primary processes of the universe, generating much of what we see around us, present in processes from Schrödinger's wave equation to the generation of new species in nature and the generation of new ideas by humans.

57.

Let us look at the term "information" a bit more closely:
> Etymologically the term information is a noun formed from the verb "to inform," which was borrowed in the 15th century from the Latin word "informare," which means 'to give form to,' 'to shape,' or 'to form.' During the Renaissance the word 'to inform' was synonymous to the word 'to instruct'. (Mark Burgin 54)

To inform (inform as a verb) is to put into form. But if something is inform (inform as a noun), it is without form. Information is formless, yet it gives rise to form. It is thus creative. We get construction through instruction. In "Passing Messages Between Disciplines," Marc Mézard says that "Complex behaviors can emerge in systems in which many "atoms"—such as real atoms, economic agents, logical variables, or neurons—locally exchange messages" (*Science* 19 Sept. 2003, 1685). Information is transmitted from one component to another through the transfer of energy, which is itself a kind of information. Electromagnetism is transmitted from electron to electron through photons. Electromagnetism is the information—the carrier of the information is the photon—the sender and the recipient of the information are the electrons. When an electron receives a high-energy photon, it leaps into a higher energy level—releasing another photon at a lower energy level when the electron drops back down to its previous energy level.

58.

"Energy" comes from the Greek, *en,* for "in", and *érgon,* for "work" or "deed," implying action, or doing. Energy is the amount of work, deeds, or actions a system can do. The energy content of the universe is the work, deeds, or actions the universe can do—to make matter, molecules, life, and human intelligence. What we call forces are also an ability to do work. Since matter is densified energy, matter can be seen as densified work, deeds, or actions. In the same way matter contains/is energy, knowledge/data/signs/text contain/carries information. Matter is similar to knowledge/data as energy is similar to information (Burgin, 62-3). If we come back to Nietzsche's Will to Power, and recognize that in physics, Power = Work/Time, we can begin to see what Nietzsche meant by his idea of power, especially if we combine his Will to Power with his Eternal Return, which is his understanding of the nature of time.

59.

Physis and *logos* are one; *nomos* and *logos* are one; *nomos* emerges from *physis*, but *nomos* is not *physis*, or vice versa; to avoid tragedy, *nomos* needs to map onto *physis* in a fractal fashion.

60.

Nature and information are one; culture and information are one; culture emerges from nature, but culture is not nature; to avoid tragedy, culture needs to map onto nature in a fractal fashion. This is a much less accurate, but much more common, way of restating the situation. Unless we understand that nature includes all levels of nature: energy, quantum physical particle-waves, molecules, and living organisms, including the biological aspects of humans. And unless we understand that culture here includes all levels of a culture: culture, art, literature, music, dance, language, economy, political structure, religions, etc. All are complex systems.

61.

Fractal geometry has the following features: fractional dimensions, a finite space surrounded by an infinite border, and self-similarity regardless of scale. The latter means that there are structures that get repeated no matter how far away we are from the figure or how closely we look at the figure. When a natural object (rather than a mathematically perfect object) is fractal, it is statistically fractal, meaning the self-similarity is precisely that: statistically self-similar. Most of the universe has fractal geometry.

62.

Harriett Hawkins calls the chaos of chaos theory "deterministic chaos," which most clearly captures what this theory says about the world. Since the time of Newton and Laplace, we have understood the universe as being deterministic. Laplace in particular pointed out that with enough information, the future was completely calculable. The Romantics, including the Existentialists and postmodernists, rebelled against this notion, seeing it as an affront to freedom. They replaced the idea of determinism with the idea of a random-chaotic universe. Others, particularly the Existentialists, recommended performing gratuitous acts in rebellion against the deterministic universe (though, if the universe were determined, these "gratuitous acts" would themselves have been predetermined). This is particularly ironic considering Heidegger in his essay "The Origin of the Work of Art" recognized the power of structures we would later recognize as fractal:

> if form is correlated with the rational and matter [content] with the irrational; if the rational is taken to be the logical and the irrational the alogical; if in addition the subject-object relation is coupled with the conceptual pair form-matter; then re-presentation has at its command a conceptual machinery that nothing is capable of withstanding. (*PLT*, 27)

Disorder is given form by information. Nietzsche rebelled against Newtonian-Laplacian determinism and Romantic notions of freedom by proposing the eternal return. Chaos theory solved this problem by showing the universe is both determined and random-chaotic, simultaneously. Every system in the universe is simultaneously ordered and disordered; determined-chaotic. This is particularly clear in the images of fractal geometry (geometrical forms with fractional dimensions—and thus more accurately represent nature), which includes the Fibonacci spiral and the Mandelbrot and Julia sets. The latter two are examples of finite volumes contained by infinitely-long borders. These borders are infinite because they fold in on themselves, creating repetitions of the same structures regardless of scale, over time. Magnify any portion of a fractal's border, and it will eventually show similar images, including images of the larger form. The borders of the Mandelbrot and Julia sets fold in on themselves because fractals are created by strange attractors which, unlike simple attractors like gravity, which only attract, have the property of both attracting and repulsing, depending on the proximity of the border to the strange attractor(s) of the system at any given moment in time. Time truly is the essence of chaos theory and fractals. In discussing fractals, we must remember (to avoid Wolfram's mistake regarding fractal geometry in *A New Kind of Science*) that, "Self-similarity comes in two flavors: exact and statistical." Exact "displays an exact repetition of patterns at different magnifications." In statistical, "the patterns don't repeat exactly; instead the statistical qualities of the patterns relate. Most of nature's patterns obey statistical self-similarity, and so do [Jackson] Pollock's paintings" (Richard Taylor, "Order in Pollock's Chaos" *Scientific American*, December 2002, 118).

63.

Gravity is a regular attractor. If we swing a pendulum, it will settle down to a point, attracted to it by gravity. A strange attractor is one that pulls an object toward it, but then pushes it away, only to attract it again, etc., creating a complex system (objects moving in time). The attractor is strange in that it is both unpredictable and exhibits both attraction and repulsion. One cannot predict when the push or pull will occur, making the system chaotic. But pushing and pulling (bifurcation) will occur, making it also predictable as a system, meaning it will resemble (but not be identical to) former states of the system. This is what makes it "deterministic chaos."

64.

Dissipative structures are how real-world (versus mathematically modeled) deterministic-chaotic and biotic systems realize themselves. A dissipative structure is a system organized through/from/because of entropy. In a closed system, entropy increases over time, and order becomes disorder. But in an open, complex system, certain elements can actually make use of entropy increase to create local decreases in entropy. Entropy increase in one place is transformed into a structure in another place, which releases energy, as entropy. A system produces the most work by being in a Maximum Entropy Production (MEP) state (Ralph Lorenz, *Science* 7 Feb. 2003, 837), or dissipative structure. So long as there is an energy source to produce work, the system retains its structure, as it transforms the energy into other forms. The sun, as it engages in fusion, increases in entropy, releasing energy in the form of radiation and light. This light is absorbed by chlorophyll in plants and is used to turn water and carbon dioxide into complex molecules (sugars) and oxygen by photosynthesis. The plant uses these complex molecules to create energy to run other systems and cycles in the plant, to grow and maintain itself, and loses energy as carbon dioxide, oxygen, water, and heat. Energy is taken in, transformed, and released. In the transformation, structure is created. These structures have emergent properties unpredictable from their constituent parts, and are capable of creating themselves from their constituent parts. The parts co-operate to create an emergent structure. A simple example of this was explained in a 17 Jan 2003 article in *Science*—rock rings in arctic areas. If one were to look at these rock rings, the first thing one would ask is, "Who came out here and organized these rocks into rings?" In this case, the answer is, "Nobody." Rocks in barren arctic and antarctic regions actually organize themselves into interconnected rings through the simple interaction of the different-sized rocks with freezing and thawing water. To get rings of rocks on a landscape, all that is needed is water, rocks, and temperatures varying enough to allow water to freeze and thaw. If rocks can organize themselves into rings, imagine what molecules with complex chemical and physical interactions could organize themselves into. But why imagine? Life is precisely the self-organization of certain molecules into dissipative structures. Every cell is a self-organized, self-generated structure made up of different parts interacting with each other. Simple physics will give you rock rings. Organic chemistry will give you life. This understanding of nature as self-organizing is related to the Greek word *physis*, "that which generates itself," from *phyein* "to produce." We have thus returned to the Heraclitean view of *physis* in understanding the cosmos as self-creating, self-organizing. Another component of complex systems is their composition of components. Each of those components, I will argue, are themselves dissipative systems, meaning each level of complexity is scalarly invariant to every other level of complexity to the extent that each level is a complex system, while at the same time the complex interactions of complex systems—continue to act as individual parts—give rise to emergent properties of the new system. Each new system is a magnitude more complex than its parts would suggest. This is due to the co-operative aspects of the system. "Co-operation is considered in a broad sense as a phenomenon that can be found in all complex, self-organizing systems" (Fuchs, 2). And not only is each system more complex as one goes up the hierarchy, but "Co-operation is itself an evolving phenomenon, during the course of its evolution new higher emergent qualities and levels of co-operation arise that can't be reduced to lower levels or qualities" (2).

65.

When one thinks of evolution, one usually thinks of evolutionary biology and Darwin's theory of natural selection. But evolution can be applied to complex systems in general, not just biology. An ordered system can become increasingly disordered until it finds itself on the borderlands of order and random-chaos. Dissipative structures achieve higher levels of order by moving into and past this borderland—while at the same time retaining some random-chaos in their new, more complex order. Ilya Prigogine applies this idea to cultural evolution too—societies remain ordered over long periods of time, then go through a random-chaotic period leading to a new order. Clare Graves further solidifies this idea, developed in Don Beck and Christopher Cowan's *Spiral Dynamics*. Examples of such evolution include the move from the archaic pre-Socratic Greeks, through the Tragic Era, to the median Platonists and Aristotleans, and the move from Medieval Europe, through the Renaissance, to the Modern Era. This kind of evolution also occurs in the evolutionary theory of punctuated equilibrium. Jay Gould observed that species tend to be evolutionarily stable over tens of thousands to millions of years, then suddenly change into (a) different species. Mutations accumulate at a steady rate without affecting the population, until a critical number of mutations accumulate, and the species rapidly changes into a new one, or ones. The rapidity of human evolution is a good example of this. I discuss biological evolution in particular because we must look to our distant genetic ancestors to understand how we became what we are. "All the genetic instructions for today's living species are built upon the modified instructions for ancestral species" (Gribbin & Cherfas, 152), and this includes humans. A similar kind of evolution can also be applied to the experience of time, as we will see in J. T. Fraser's *umwelt* theory of time, which he calls the evolutionary and the hierarchical theory of time. Now, while I have mentioned evolution in relation to dissipative structures, I should point out that one thing evolution is not is an explanation of how one can get more complexity. Evolution is simply change, and the theories of evolution explain how those changes get selected for and against. There is no such thing as "progress" in a teleological sense in Darwinian evolution. If one wants an explanation of how one gets increased complexity—the definition of progress I use—one has to look instead to systems, dissipative structures theory, emergence, and game theory.

66.

Game theory uncovers the rules of complex systems. It shows that complex systems—particularly complex social systems and biological systems, as John Maynard Smith has shown—came be understood through game theory, meaning they follow rules. In postulating that games have rules, game theory goes against certain anarchist views that insist on opposing the very concept of rules. They see rules as limiting, preventing freedom. They do not understand that it is the very presence of rules that give us "degrees of freedom." Nietzsche points out in *Beyond Good and Evil* that rules are absolutely necessary, for every form of morality and art has used and needed rules.

> What is essential and inestimable in every morality is that it constitutes a long compulsion: to understand Stoicism or Port-Royal or Puritanism, one should recall the compulsion under which every language so far has achieved strength and freedom—the metrical compulsion of rhyme and rhythm. (188)

He then goes so far as to say that "all there is or has been on earth of freedom, subtlety, boldness, dance, and masterly sureness, whether in thought itself or in government, or in rhetoric and persuasion, in the arts just as in ethics" developed only because of rules—and that the use of rules lies in nature itself, that rules are natural. It is through living by rules that we make it "worth while to live on earth; for example, virtue, art, music, dance, reason, spirituality." Nietzsche rejects living without rules. But which rules? The tacit question asked by game theory is: "what rules make for the best games?" But it also asks: "what rules would evolve to ensure survival of the game?"—whether that game is a species or a ritual, an economic system or a work of literature. Perhaps game theory, broadly applied, can help us understand the source of rules, from the Laws of Physics to the rules of grammar. It shows that more rules are needed for more complex games. Only a few are needed at the quantum level, but with each movement up in complexity, more rules emerge—and are needed—until one gets to complex human social systems, which need thousands of rules. And it shows how necessary rules are if one is going to have any sort of game at all. It is the existence of rules that give us freedom—making us more creative, often far more creative than we are otherwise. Many good rules (note the word "good" here—it is not the number of rules so much as the kind, those that generate more moves, not less) give us many more degrees of freedom. Chess is a better, more complex game, with many more degrees of freedom, than checkers, though both are played on the same board. It is better because more complex. Complexity gives us more freedom. So game theory too fits into the theory I am proposing—no matter the scale, rules are necessary, but the more complex the system, the more, and more complex, rules that are necessary.

67.

Rhythms and patterns are fundamental to the world. Rhythms and patterns are rules found at every scale, and they are the very essence of fractal geometry. Patterns are spatially organized. They are arrangements of forms and/or colors, which is to say, they are designs. In literature, they create the motif. They are imitations or copies, which comes from the etymology of "pattern," derived from the Old French, *patron*, since a client would copy his patron. The patron provided a model for the client to follow—the patron would act as the ideal, example, exemplar. Thus there is a connection between patterns and ethics, and between beauty and ethics, since patterns are part of beauty. Rhythms, from the Greek *rhein*, to flow, are temporally organized. Rhythms are regular repetitions over time, typically of sound. Nonetheless, there is a connection to pattern: in a work of art a rhythm holds the parts together to create a harmonious whole through the repetition of form and/or color. Since rhyme come from the Greek *rhythmós*, we see that rhyme is a kind of rhythm. There is also an ethical component to rhythm, since a rhythm in biology is a pattern of involuntary behavior or action that occurs regularly and periodically. In the realm of behavior, rhythms are involuntary actions, whereas patterns are consciously followed. A rhythm is meant to carry us along, to help our actions flow, while patterns help create meaning by making us more conscious.

68.

A fundamental feature of bios is novelty. "Novelty is an empirical measure of creativity that indicates faster and larger variation than generated by random change" (*Bios*, 1). Another feature of a biotic time series is that it gives a mandala image when you calculate "the sine and cosine of each interval. Sine and cosine are complementary opposites, waxing and waning out of phase" (2). Thus, mandalas are an archetype and, with bios, "We have now found this cultural archetype in a biological process, just as the divine proportion [phi], first discovered by artists, was later found in botanical and anatomical structures" (4). Physically, "creativity defines bios and differentiates it from chaos. Unpredictability characterizes chaos; *novelty characterizes bios*" (7). Further:

> Biotic processes generate irregular forms that appear random or chaotic, but they display complexes (forms with short life span), novelty, diversity, nonrandom arrangement, and the coexistence of simple and complex patterns (indicating the causal generation of complexity. In contrast, mechanical, periodic, and chaotic processes maintain their form unchanged. (7)

Biotic processes involve creative interactions of opposites, generating creative phenomena. As Sabelli observes repeatedly, creation must necessarily precede destruction, and bios creates order. Bios also shows us that "complex processes feedback into their simple environment and control it through their greater informational content. Simple processes have priority, and complex processes have supremacy" (8). Which brings us around to independently affirming J.T. Fraser's *umwelt* theory of time and emergentist evolution and Graves-Beck-Cowan's Spiral Dynamics theory, both of which state that the simple has priority and the complex has superiority. No life without molecules; no molecules without atoms and quantum physics.

69.

An object is not "in" space and time as if space and time was a container the object was placed in. Every object in the universe constitutes spacetime—spacetime becomes densified, more concentrated in one place with emergence into new levels of greater complexity; but still, that object constitutes spacetime. This is why each new level of complexity experiences time in different ways—complexity comes about from recursion and interaction, and thus time experience becomes increasingly recursive and interactive.

70.

The universe is neither analog-wave nor digital-particle, but both simultaneously. This is in agreement with quantum physics, which explains the fundamental nature of matter as being both particle and wave. This paradoxical structure of quantum "objects" is reflected in each hierarchical level in the universe, whether quantum, molecular/macrophysical, biological, in social animals, in humans, or in art and technology (this last one is really more a product of human thought, and as such more complex, but not really a truly emergent, exponential level of complexity, since they are not typically systems, but part and product of the human-intelligence system). Each of these, when they are systems, are both digital and analog simultaneously. That is how a system is able to be a system—it has parts, working as individual parts, but working together with other parts to create a system that has emergent properties. One has to have quantum physics and the atom to have chemistry, and one has to have organic chemistry to have biology, and one has to have animals to have social animals, and one has to have social animals to develop human-type intelligence, which is necessary to create art, philosophy, and technology, which itself acts in feedback fashion to influence human-type intelligence.

71.

In *Nonzero* Robert Wright shows that density leads to complexity (he is talking about density of human population leading to greater complexity, but this applies, as we shall see, on all scales. As John T. Bonner points out, "The first major step toward culture is the centralization of the nervous system and the formation of a brain" (*The Evolution of Culture in Animals*, 38), which allows neurons to become denser, so we can see how denser energy (work) leads to more complex matter. "Information is what synchronizes the parts of the whole and keeps them in touch with each other as they collectively resist disruption and decay"; that is, "Information is a structured form of matter or energy whose generic function is to sustain and protect structure" (Wright, 250). Information theory says information is a kind of entropy—its inverse—and entropy is the loss of energy and order from a system. If a system loses energy, it is giving off information about that system. We, and all other systems in the universe, pick up that energy as information, which tells us something about the state and nature of those systems. Energy taken in is first stored in memory, then turned into work. Energy given off is given off as information, which itself can be turned into work by another system. In regards to energy and its connection to a system's rules (connecting information to game theory), Heidegger points out that

> The boundary in the Greek sense does not block off; rather, being itself brought forth, it first brings to its radiance what is present. Boundary sets free into the unconcealed; by its contour in the Greek light the mountain stands in its towering and repose. The boundary that fixes and consolidates is in this repose—repose in the fullness of motion—all this holds of the work in the Greek sense of *ergon*; this work's "being" is *energeia*. (*PLT*, 83)

These boundaries are rules that give energy to a system to create its work. These boundaries (rules) are therefore creative, not restrictive. Heraclitus recognized that *physis/logos*, as exemplified by Apollo, "neither declares nor conceals, but gives a sign" (K. XXXIII). Information is that sign.

72.

Physis as *logos* gives signs (information) about itself, neither hiding nor uncovering itself. This is why *physis* and *logos* need interpretation. Interpretation of the *logos* is independent of and prior to the existence of human beings. *Physis* carries on a discourse with itself—one way of which is through human interpretation. The universe knows itself better because we are here to know it. By understanding *physis* as *logos*, and vice versa, as Heraclitus did, we can see how understanding comes about through the interpretation of signs. We get a connection between community and communication in the fact that *logos* itself is both communication, and means "collection," a collection or gathering together of people being a community, from *legein*—"to collect", "to gather." When we read a book, we take in the content as information. As we read, our brains convert that information into work, putting elements of the work into conceptual slots, organizing the information and storing it, changing the way we think, conceive, organize. The brain accumulates information until you feel the compulsion to create. The information is put to work in the brain, and the brain puts it out as new information. This information is not "pure" information—nor would we want it to be, as the equations of information theory (which is the same as that for entropy: $S = k \log W$, where S = entropy, k = Boltzmann's constant, and W = number of ways the system can be arranged; low W = low S; as W approaches infinity, so does S) show that if it were pure (zero noise; $S = 0$), information would equal randomness ($W = 0$; or, system can be arranged zero ways). To have information, one must have noise, or ambiguity. Ambiguity in a text does not prevent one from getting meaning from the text; it helps give a text more meaning, particularly in the use of redundancy, or repetition, to reduce noise. One of the jobs of theorists, scholars and philosophers is to uncover the meaning(s) of a work, which is to say, its underlying (or overlying) patterns, including the works of *physis, logos, nomos,* etc. In the same way, the job of the scientist is to uncover the information of *physis*. We can also apply information theory to every other scale of the complex nested hierarchy of systems that is the universe, though the complexity and kind of information increases with the complexity of the systems.

73.

The genome and the brain are two kinds of information-processing and -creating engines. There is also a relationship between the two that highlights the relationship between each level of the world's complex nested hierarchy. What we have with the genes' relation to the brain is:

> one information processing machine (the genome) has spawned another (the brain). Furthermore it has created a machine that can process information in new and different ways, the most striking of which is the difference in the rate of processing. The slow genome has, over millions of years, given birth to the rapid brain. (Bonner, 30)

There is a parallel between the two in that they are both dynamic parallel-processing systems:

> The brain processes thoughts, movements, immediate reactions to the environment, in sum all the activities we associate with animals. The genome processes genes by replication, and the genes are responsible for making specific proteins that in turn are required to build the structure of the organism through its entire life cycle. The basic similarity between the two is that they both take in, store, and give out information; the difference between the two is not only that the information differs, but that they are on a different time scale. Reactions of the brain and the nervous system are rapid, while those of the genome are, by comparison, exceedingly slow. (Bonner, 30)

The rapidity of the brain's work relative to the work done by the genome has led people to consider the brain a system which could not possibly be related to the genome that codes for it. This would be like saying that because atoms process information even slower than the genomes they form, that the two could not possibly be related. Atoms, the genetic system, and the brain are closely related in that all three are dissipative systems, and all dissipative systems "take in, store, and give out information." Greater complexity gives rise to greater speed of information processing—especially of more complex information.

74.

Information is what is taken in by a system. Entropy is what is given off by a system. A dissipative structure is the kind of structure that takes in information, processes and stores that information, and gives off that information as entropy. Consider the following scenario: the sun gives off light as entropy. Light is picked up by chlorophyll as information, and is transformed into sugars, which are stored and used for energy use in the plant. A mouse comes along and eats one of the seeds of the plants, taking in the information from the seed, transforming it into sugars and amino acids for the mouse. Some of this information is released as entropic heat, which is detected as information by the heat sensors in the pits of pit vipers such as rattlesnakes. The snake uses the information to find and kill and eat the mouse, converting the mouse into information stored and transformed by the snake. The snake in turn uses some of this energy from the mouse to use its rattle to transmit information to those animals that would like to eat it that it is poisonous. So here we see information transfer among living species, beginning with information given off by the fusion process of the sun.

75.

On the quantum physics level of reality, fermions (matter) exchange information using bosons. Electrons and electron neutrinos use photons to transmit electromagnetism. Quarks exchange gluons to transmit the strong nuclear force to each other, thus creating atomic nuclei.

76.

It is incorrect to think of quantum "objects" as particles, though that is usually how we think of them. It is equally incorrect to think of them as waves, though that is usually how we think of at least some of them (photons, for example). In Shimon Malin's factually accurate but wildly incorrect interpretation of quantum physics (especially as it relates to philosophy), *Nature Loves to Hide*, we learn what is actually the case in quantum physics. What we have are inform waves that, in interacting with other inform waves, create a zillion little wavelets that, interacting with each other using system dynamics, emerge into decohered particle-form, which then immediately dissipates back into wave form. Each electron is a dissipative system, it's own wave function providing the information necessary to inform the electron and emerge into the particle-form, and dissipate again into its own wave function. It provides food for itself.

77.

But let us return to Malin for a moment, for it is important to show why Malin is wrong about his conclusions. Malin is trying to do little more than reintroduce transcendentalism through quantum physics. Despite this, I appreciate him, because he gives a lot of very interesting and important information, and he is wrong in very interesting ways. He is wrong in such a way that, because he combines it with good information, I can see precisely why he's wrong as well as what would be a better interpretation. For example, he observes that electrons are only waves of potential until they are actualized in the collapse of the wave function. He then goes on to say that all of this potential is outside of spacetime, and that it is atemporal. He then suggests that this "place" where it is making its decisions to actualize (or not) is in some transcendental space. However, when he points out that waves of potentialities "interfere, reinforcing or weakening each other, as the case may be. Here too, a few interfering waves can separate and recombine in clear patterns; but when a wave pattern breaks up into a zillion little waves, a recombination is no longer possible, and a collapse occurs" (128), I came to realize that the actualized, particle-form of the electron is actually an emergent property of these "zillion little waves" interacting with each other in a complex system. As the analog potentiality waves begin to break up and become more and more complex and more and more digital, they suddenly are capable of interacting in a more complex way, and give rise to the digital form of the electron-particle in the wave collapse. The electron is then capable of dissipating back out into its wave form. Thus, an electron is truly self-assembling, as it assembles itself from itself. Further, an electron can only do this because it is in an ecosystem of other particle-waves. For example, imagine a universe that only contained a single electron -- it would never collapse, it would always retain its wave function, because it does not have the interference of the rest of the universe to complexify its wave form for it to collapse. Thus it would never emerge into particle-form. It would be like a human never surrounded by other humans -- such a person could never emerge into having a human mind. Humans have to "observe" each other into becoming fully human. Similarly, quantum particles get observed into collapsing by complexifying the waveform until there are enough to create the emergent collapse of the wave function.

78.

Consider the statement that "an electron is a field of potentialities." Let's translate that into a few extended definitions: "an electron is a region of space in which a physical property has a determinable value at every point of the inherent ability (having sufficient power, or work(energy)/time) for growth, development, or realization." Therefore, an electron is not _in_ spacetime, but constitutes spacetime—a part of spacetime capable of collapsing to an object "in" spacetime. Every quantum particle-wave constitutes spacetime when it is in its waveform. Or, quite frankly, when it is in its particle form. Therefore spacetime itself contains the potentiality for collapse of the waveform. The waveform of spacetime is thus the analog aspect of the universe, while the collapse of the waveform into particles is its digital aspect. This explanation is clarified even more when we understand everything as information—that that which is inform gives form by informing.

79.

Malin makes a big deal out of the fact that "potentiality" cannot be "located" anywhere (120). But if we really understand $E=mc^2$, we realize that all energy is potential mass, and all mass is potential energy. Each contains the potential of the other. But potential is useless unless realized. What matters is whether or not that potential is actualized. Malin is making the mistake of thinking that because the probability of having two bombs on a plane is much lower than having one, that he should carry a bomb with him every time he boards a plane, in order to reduce the chance of having a bomb on the plane someone wants to detonate. He has in fact affected nothing (Paulos, *Innumeracy)*. Potentiality is the same. Uranium can sit around, perfectly stable, for millennia. Any given uranium atom is potentially radioactive—the ones that matter for radioactivity are the ones that actually undergo radioactive decay. Further, potentiality simply implies ability. Uranium has the potential to decay because it is able to decay. The potential lies in the qualities of the entity—it is neither mystical nor transcendental. It is an abstract idea, and thus a creation of the human mind—useful in explaining, for example, why something that was not undergoing radioactive decay suddenly decays. Thus, we must beware of mistaking concepts, which are creations of human minds, for having external reality in their conceptual form. This is the mistake of those who believe in the Forms as preceding perception.

80.

We cannot locate "freedom" anywhere "out there" in the real, physical world—so does that mean it only has transcendental reality, that it is one of the Forms? Concepts are created by comparing similar, yet unlike, things and removing all the differences. "Freedom" is one of those things we conceptualize so that, because it includes so many unlike things, it becomes pure abstraction. In every case, freedom involves choices. The more choices we have, the more freedom we have.

81.

"Rules are a form of redundancy" (Jeremy Campbell, *Grammatical Man*, 69). Another way of wording it: "Redundancy is essentially a constraint" (68), meaning they reduce "the number of ways in which the various parts of a system can be arranged" (68). Without redundancy, there would be no complexity, since "The more complex the system, the more likely it is that one of its parts will malfunction. Redundancy is a means of keeping the system running in the presence of malfunction" (73). Redundancy allows a system to be maintained between total order and total randomness — thus giving the system true freedom. Anarchy is not freedom. It is as restrictive as complete order. Artists and AI researchers alike must hear this message: art and creativity only come out of the greatest discipline. Discipline is what turns randomness into complexity and makes a system truly free in its expression.

82.

The word "potential" comes from the Late Latin word *potentialis*, which means "powerful," from the Latin *potentia*, meaning "power." It is derived from the Latin word *posse*, meaning "to be able," and is a composite word from *potis*, meaning "able", and *esse*, meaning "to be." So to have potential is simply to be able to do something, to have the power to be able to be. Malin will have us believe that ability does not exist anywhere; I would suggest that ability lies within the entity itself. If I were to say that I have the potential to write a publishable short story, if we do not take the term to absurd heights and say that everyone who is literate has the potential to write a short story, as true as that may be, then what I am saying is that I am able to write a short story. How do we know that I am able to write a short story? Because in the past I have in fact written short stories that have in fact been published. Thus we know that the potential for me to write another publishable short story exists. However, suppose we say that someone else has such potential, but nothing that person ever writes actually gets published. Did that person in fact have this potential? I would argue that he does not, since he was never actually able to get his stories published. If we push the etymology of "potential" back further, to the proto-Indo-European, *poti-*, we see that it means "powerful" or "Lord", and that the word gives rise to the words "possess", "possible", and "potent." If something has potential, it possesses something that makes action possible. Such action is "potent," that is, it results in growth. Thus potential is something possessed by

Physis

the entity to make it possible for it to act. Meaning it is something that exists in spacetime.

83.

Potential is made actual through information. "Information is in essence a theory about making the possible actual. It sets an event which does happen in the context of other and different events which only might have happened, so that potential and actual are related" (Campbell, 269). Possibility is realized through interactions. To get decoherence, which Seth Lloyd defines as "the process by which the environment destroys the wavelike nature of things by getting information about a quantum system" (*Programming the Universe*, 108), all you need "is for some system, no matter how small, to get information about the position of the particle" (108).

84.

What we refer to as mechanistic or macro- physics is nothing more than the behaviors emergent from non-living chemistry and/or multi-atomic clusters. This is why mechanistic physics is explainable by such simple mathematics, as emergent properties have simpler explanations than if we explained the system at a lower hierarchical level of reality. Atoms interacting on a large scale give rise to mechanistic physics. Quantum physics bifurcates and emerges into either 1) fusion physics, or 2) chemistry. Of course, any chemistry more complex than hydrogen or lithium chemistry is impossible prior to fusion physics, since it is by fusion physics in stars that we get complex atoms able to create complex chemistry. This is nonetheless a bifurcation of quantum physics precisely because each are very different things from the standpoint of quantum physics. Next, chemistry bifurcates to give emergence to either 1) non-system objects, which are explainable by mechanistic physics, or 2) system objects, which exist as either a) nonliving objects, or b) living objects, which are explainable by biology. Living objects first bifurcate into eubacteria and archaebacteria, and then these two combine to create eukaryotes, creating a split between prokaryotes and eukaryotes. The eukaryotes split to become single-celled and polycellular organisms. Some of the polycellular organisms split off into fungi and plants (another bifurcation), and some polycellular organisms retained some single-cells, in a sense combining to create multicellular organisms, such as sponges, which are both multicellular and have free cells swimming around among the connected cells, creating more complex communication among the cells throughout the organism—thus dividing animals into simple and complex animals. We get further bifurcations into the branch that eventually gave rise to animals with exoskeletons, and the branch that eventually gave rise to animals with endoskeletons, which gave rise to the fishes, which bifurcated into fish and amphibians, which bifurcated into amphibians and reptiles, which bifurcated into dinosaurs and reptiles, which bifurcated into reptiles and mammals, which bifurcated into nonsocial and social mammals, which bifurcated into nonprimates and primates, which bifurcated into monkeys and apes, which bifurcated into quadrupedal and bipedal apes, which gave rise to humans when mating songs bifurcated into music and language. This is of course a very simplified version, and it is also anthropocentric (I left out that the dinosaurs bifurcated into lizard-hipped and bird-hipped dinosaurs, which bifurcated into bird-hipped dinosaurs and birds, for example). Some of these bifurcations have gone extinct, such as the dinosaurs, but the bifurcation was there, nonetheless.

85.

Photons are electromagnetic particle-waves. So are electrons. One is a boson, the other is a fermion. Electrons jump from one orbit to another in atoms and are not found in the spaces in between—this is known as quantum jumping, a kind of quantum tunneling. Electrons act continuously, as waves, and are spread out. Thus, they act as both particles and as waves. This is part of the digital-analog problem—a paradox that explains the actions and interactions of all systems.

86.

I mentioned earlier that in the arctic and antarctic are self-organized rock rings. These rocks, interacting with only water and fluctuating temperatures (freezing-nonfreezing), are able to organize themselves into rings, with larger rocks in the rings, and smaller rocks toward the middle (in a power law distribution, of course). These rocks are digital entities acting in continuous fashion in pulses (digital-analog) to create patterns—they are pieces acting together, in concert. However, when people first saw these loops and rings of rocks, the first thing that came to their minds was "who came out here and did this?" That is, people first thought some kind of intelligence was needed to create this kind of order. The theory of dissipative structures showed scientists, however, that nobody was needed to organize the rocks. The rocks could organize themselves, since they were part of a dynamic system. Whenever we see something orderly and organized, we automatically assume there must be an intelligence behind it. We have this instinct because it is true: intelligence is needed to create order. However, it is not the kind of intelligence we typically think of. The rock rings give the appearance of intelligence because the system is intelligent in a limited way. One good definition of intelligence may be: the ability to actively turn randomness into order. Or, the ability to create patterns. If we accept this definition of intelligence, all dynamic emergent systems are intelligent, since they all turn randomness (entropy) into order (information). This order is patterned because the universe is patterned all the way down, from the wave functions of quantum entities up to rhythmic, rhyming poetry. This is in agreement with Jeff Hawkins' definition of intelligence in *On Intelligence*, and Ray Kurzweil saying the brain is a "digital controlled analog electrochemical process" (www.kurzweilai.net/articles/art0134.html?printable=1). Thus, the creation of a decohered electron from its wave function is an intelligent process, even if it is the lowest form of intelligence. And the interactions of quarks and electrons to create atoms is a kind of intelligent process as well, since an atom is a dissipative structure. And onward and upward.

Physis

87.

Another definition of intelligence is the other side of what I stated above, and it is the main thesis of Jeff Hawkins' book: that intelligence is the recognition of patterns and, thus, the ability to make accurate predictions. Do the rock ring systems recognize patterns and make accurate predictions? Hardly. However, animal- and human-type intelligence are not the only kinds of intelligence—though it is the kind Hawkins is most interested in discussing. To more fully understand the differences among these kinds of systems, we have to consider the temporal experience of each kinds. Here we get into the philosophy of J.T. Fraser.

88.

Before I get into Fraser, let me pause to say that I may have given the impression that everything is a dissipative structure. A crystal quartz is not a dissipative structure. It is orderly and relatively stagnant and deterministic at the molecular level. All those things that are deterministic, that are calculable using mechanistic physics, are not dissipative structures. Mechanistic physics uses linear dynamics. Dissipative structures and complex systems use nonlinear dynamics. A quartz's components, the atoms, may be dissipative structures, but the structure of a quartz is not. There are emergent dead ends. When we talk about emergent systems, we have to differentiate between those that can and cannot themselves give rise to new emergent systems. Even though some fish were able to emerge into amphibians, not all fish were able to do so. All the fish are emergent as fish, but only the lobe-finned fish had the physiological features to emerge into amphibians. A sardine is in a sense an emergent dead-end. It can and will only evolve into another kind of fish, should any given species of sardine happen to bifurcate.

89.

What is time? A nonsensical question. Time is not a "what." Furthermore, "is" implies equation, unchanging being, permanence. If anything is opposite these, it is time (but time is not an "anything," including an "it"—and I've used "is" twice again). Is time a measurement of change? A second is a measure of time, the same as a centimeter is a measure of length—but we would not say length is a measure of space, per se. The problem is that "time" is a noun, which carries with it the idea of something unchanging.

90.

The past consists of everything that happened. Science and history are both the same field in this sense: both are concerned with uncovering what happened. The past has the convenience of being very orderly, of being dead. That makes it relatively easy to study. At the same time, because it is dead and orderly, it can give us very incorrect ideas about the nature of things. How, rather, can we come to understand and study living things?

91.

The present is now, on the cusp of the immediate past and of the future. It is built on the dead wood of the past.

92.

There is no such thing as "future." This is an imprecise statement. Nothing is in the future. No thing is in the future. There is no such thing as future. It has no where or when or objectivity (or subjectivity). The future is void. It is the void into which the present is pushed by the past. One cannot know anything about the future, just as we can know nothing about nothing (by definition). And what of prediction (which is, after all, a kind of knowledge)? We can make more or less accurate predictions about what will happen in a future present: this is the heart of intelligence and is another important aspect of science. Deterministic things will act deterministically. Probabilistic things will act probabilistically. And we can make predictions accordingly. But still, we have to wait for it to become present for us to *know* if we are correct.

93.

Or, there are "futures." All of our potential actions create potential futures, all of which are simultaneously actualized and not actualized. The future branches out into trillions of branches, and each of our decisions trims the branches, removing all potential futures that could have been realized had we made another decision, another action. A tragic situation. Every decision, every action we take, alters what will happen, removes branches from the future, makes extinct every other possibility—and, with it, every potential life that could have been born had we made that decision. But then, if we had made that other decision, then all futures connected to the one we did make would have vanished. And even if you choose not to decide, you still have made a choice, and those sets of futures have been selected for. There is no getting out of the tragic situation of time. All one can do is try to make the best decisions one can—and even when we do that, we cannot tell what those futures will bring. Frederick Turner suggests in *The Culture of Hope* that futures could speak to us, encouraging us to take paths toward better futures. But are we listening? Can humans even hear what is being said? Do they have the complexity (which is to say, the level of spacetime folding) needed to hear the future? Or is that what the next level of complexity will bring?

94.

J.T. Fraser has developed a theory of time remarkable in its ability to explain each level of emergent complexity through its experience of time. Time, for Fraser, changes over time—or at least the experience of time changes over time. There is one sense in which time is the simple measure of the expansion of the universe—as space expands, time presses forward as well. We must not forget that we are always talking about spacetime—space and time as one. There is another sense of time wherein time is understood as the measurement of movement relative to other things in the universe—or perhaps to the background of the universe itself. Scientifically, we have several versions of time. We have Newton's time, T, of mechanistic physics. Time, T, does not have a particular direction, which has caused many physicists, including Einstein, to postulate that time itself is an illusion. This is perhaps aided by Einstein's formulation of time in relativity, where time experience is relative to the speed at which you are traveling. Thus, if you could travel at the speed of light, time would stop for you—you would become atemporal. This implies that the time-experience of any-thing traveling at the speed of light experiences time atemporally—there is no time-experience for a photon, for example. While this led Einstein and others to postulate that time is an illusion, this caused Fraser to postulate a view of time that was able to integrate and explain how Einstein and Newton could both be right, and how we humans could also be right in insisting that time does in fact exist, and that it goes in a preferred direction, due to our experiencing the passage of time. Clearly, we are born, live, grow older, and die. And this always happens in a particular sequence.

95.

In Fraser's theory of time, new levels of reality have different time experiences. Time at the moment of the Big Bang, of pure energy, is atemporality. At the level of quantum physics, which pure energy becomes as the universe cools from the initial conditions of the big bang, time-experience turns probabilistic (this Fraser calls the prototemporal). As we get fluctuation between particle- and wave-form, we get fluctuation between atemporality and time experience. This is perhaps why quantum particle-waves in fact propagate in particular directions rather than truly randomly. As atoms interact to create macro-atomic forms, including molecules and fusion furnaces (stars), we get chemistry/mechanistic physics—and thus Newton's time, T (which Fraser calls the eotemporal). Such objects experience time as directionless. To the extent, though, that these atoms are in systems, there is an experience of a forward direction of time (here I am adding to Fraser)—stars do burn up their fuel, meaning time has a particular direction for the star, making the star burn itself out, hydrogen turned into helium and other, more complex atoms, in a particular temporal direction. With the evolution of life (the biotemporal), we begin to get a stronger experience of the forward direction of time. Living organisms are born, live, grow old, and die. Further, for a single cell to locate food, there has to be an ability to know which way to go—there has to be planning (even in the very limited sense that a single cell can "plan" to go in a particular direction), which means there has to be at least a slight forward direction of time experience. As animals gained more complexity, including more complex behaviors and brains, the temporal horizon slowly expanded. Then, with the evolution of humans—and of language—we get the evolution of a very long temporal experience (the nootemporal). We are aware of all 14 billion years the universe has existed. We are aware of time before we existed. We are also aware of time after we die—that the universe will continue on beyond our own personal lives. We even make predictions of what the universe will look like in some future time, billions of years from now. And not only that, we can do more with the physical time we have—in a real sense, we have more time in the time we have. The advent of the Internet has shown us that we are able to integrate more information more quickly—we are speeding things up, and we are keeping up with the speed. We can integrate more information more quickly than anything else on earth. Because we can experience things faster, we can do more things in less time—meaning we do in fact have more time available to us than any of the other levels below us. Of course, any living organism can also experience and integrate information much faster than the level below it—chemistry/mechanistic physics. And this level does so faster than the level of quantum physics. Thus, time experience evolves.

96.

One could view the evolution of time experience as each level experiencing time more and more accurately. It could be how the universe comes to know itself, by evolving newer levels of complexity increasingly capable of experiencing more of the universe. Of course, part of learning about the universe is then coming to learn about each of these new levels of complexity. And, at our level, that would mean ourselves as well. And new things are always coming into existence—change and creativity do not stop once we came into being. In fact, we only add to the complexity, with our art and technology. We have learned a great deal about *physis* through reductionist science—and will undoubtedly continue to do so. But that is also only half of the equation in understanding *physis*. We have been, and must continue more and more, investigating emergence as well—complex systems, and how they work, and the ways that a system is more than the sum of its parts. We must have emergentist science too.

97.

How does complexity emerge? "The theory of quantum mechanics gives rise to large-scale structure because of its intrinsically probabilistic nature. Counter-intuitive as it may seem, quantum mechanics produces detail and structure because it is inherently uncertain" (Lloyd, 49). Chances are that a number of waves will become particles in a small region, which, because particles have mass, will gravitationally bend space there, making it more likely that more particles will accumulate there. The probabilistic distribution of particle-waves resulted in tiny variations amplified through butterfly effects into large-scale structure (49-50). Since "gravity responds to the presence of energy, where the energy density is higher, the fabric of spacetime begins to curve a little more" (195). This "gravitational clumping supplied the raw material necessary for generating complexity. As matter clumps together, the energy that matter contains becomes available for use" (199). Further, the same interactions that increase entropy "make quantum objects behave in a more classical way" (108), so entropy resulted in the production of order in the universe through decoherence and gravitation. In other words, "information tells space how to curve; and space tells information where to go" (174). As a result of these random inputs from quantum fluctuations in combination with the laws of quantum mechanics, we get "a universe with a mix of order and randomness, in which complex systems arise naturally from simple origins" (185).

98.

Emergence is the entering into new levels of greater complexity. The word "complex" means "folded"—thus the universe, as it becomes more complex, becomes more folded. Spacetime becomes more folded as it becomes more complex—thus, it becomes increasingly fractal in its geometry. As spacetime becomes more folded, more spacetime comes into more contact with more spacetime. This will affect the way space and time act and interact. As this happens, the way spacetime acts and interacts will change. For one, interactions will speed up. This idea is complementary with the previous idea that we are coming to know the nature of spacetime more accurately with emergence into each new level, as each new level also has to come to know itself, and is as much a part of spacetime as lower levels. Each new level is just a more folded version of spacetime.

99.

With emergence into each new level, those new levels are able to use more and different kinds of information and energy not available to the levels below them—much of which is only available once that level evolved (we can process and use language—which was only available to process once language emerged). Thus, it is illegitimate to say that just because one level does something, that it is relevant to all other levels. Einstein's discoveries about relativity are useless in the realm of ethics, which is a different hierarchical level of reality than physics. At the same time, each higher level is made of the levels below it—and there is a common thread, a diaphysics, that seems to unite them. Humans are a nested hierarchy of diaphysically-similar complex systems, meaning we have fractal self-similarity. We contain the biological, which contains the molecular, which contains the quantum, which contains energy. And each level transmits up certain aspects of its reality. There is a randomness to quantum physics' probabilistic experience. And chaos theory expresses both randomness and probability in a deterministic fashion—as we see in fractals. And each of these are expressed in organisms. The purposeful behavior of organisms is found too in humans, though we supplement it with symbolic and concrete goals. This feed-forward of each level's reality seems to be strongest precisely in those systems that have emerged into higher systems. Rocks have less complex fractal geometry than cells, precisely because a rock is a systems dead-end.

100.

The most complex forms of order are biotic dissipative structures. They are ordered precisely because they are disordered. They have form precisely because they are informed by the inform. They have permanence only when constantly changing. They contain within them the potential to creatively bifurcate. When change stops, we get death, and when we get death, we get entropy—change without permanence, form, or order. Too many make the mistake of thinking that the second form of change is the only form of change.

This is the origin of myths of unchanging permanence, since something has to give order to the disorder. But nothing is without change. We have even discovered that the so-called constants of quantum physics are in fact not constant. The electroweak force has changed. Planck's constant has changed over time. So has the speed of light. Probably all the constants of nature have changed—making them inconstant constants. The rules of nature themselves evolve, aside from the emergence of new rules with each new emergent level. What we need to investigate is how the rules become formulated in the first place, and the nature of their evolution. Perhaps our universe is a perfect place for life and intelligence because it evolved to become such a place, each lower level stabilizing because of and as each new level emerged.

101.

Does this mean all things are inconstant, that we therefore cannot know anything? Well, it does mean all things are inconstant—though the rate of change may be so slow that it is irrelevant for any practical purpose, such as modern-day measurements. And the modern-day measurements can form the baseline for measuring past change—which is no more arbitrary than using the meter as our standard of measuring length. And as for our being able to know anything, it only means that we cannot know anything if we only consider knowledge as knowledge of those things which are unchanging. In which case, knowledge is indeed impossible, since there is no unchanging reality. Such demands on knowledge are also made by those who think knowledge must be absolute (this is really the same thing as insisting on an unchanging reality)—meaning if our knowledge is not absolute, then we do not have knowledge. Again, if we must define knowledge this way, then we cannot, by definition, have knowledge of anything. I, however, do not make such unreasonable demands of knowledge—knowledge is better understood as increasing approximations of accuracy. We cannot ever reach "the truth" of anything—we can only come closer and further away, in a fractal orbit around the strange attractor of truth. This does not mean that there are not forms of knowing that do not approach so near absolute that they might as well be absolute. If I step off a building, I will fall. If I step in front of a speeding car, it will run me over, and not just pass right through me. These things I know—and anyone who does not know these things will succumb to natural selection. But this is because the rules of physics are not going to change in any of our lifetimes. We can know what the rules of physics are now with high accuracy—and perhaps know too what they have been. We can even figure out the rate of change so we could predict what the rules of physics will look like in the future, and even predict how that could affect what the universe is like. Of course, prediction is not knowledge, per se, but it can give us a highly informed guess that, if we could be around long enough to see if we were right, could be confirmed, and thus turned into knowledge—making us more certain of our predictions for future change. As far as this goes, the universe appears to be becoming more solid, not less; more real, not less; more ordered, not less. Increasing complexity means more solid, more real, more ordered. The universe has become more and more complex over its entire history. My prediction is that it will continue to become even more complex.

102.

This correlation of certainty with knowledge is part of Malin's philosophy as well. Malin mistakes "utter certainty"—especially that associated with insight—with Truth. He gives examples of successful insights, but he leaves out all of the unsuccessful "insights" that have happened to innumerable people throughout world history. One can have utter certainty of a falsehood. This is why Malin—and many others—are incorrect when they associate certainty with knowledge and truth. The "flash of insight" gives one a feeling of eternity—Nietzsche talks about this at length, this moment of eternity. But Malin mistakes eternity for atemporality. They are, in fact, opposites. Eternity is time condensation—it is the experience of more time. In the end, Malin sees Truth as atemporal—he is mistaking both the experience of light at the speed of light and math as Truth (122-3). He thinks we can have certainty, which is Truth, which he says is atemporal. What Malin is proposing is a very old, not a new, paradigm. It mistakes the simplest of things for the most complex. Ultimately, this is why Malin suggests that there must be an "other" reality that is outside of time and more real than this reality. However, the instaneity he speaks of as atemporal would only be so at the speed of light, since no time passes at the speed of light. If light is propagating spacetime and yet experiences atemporality, the photon is still spacetime, even if its experience of it is as atemporality. It is not outside of spacetime, but simply has a different experience of it, since it is propagating at the speed of light. The same would be true of anything propagating as a wave. If an electron is a particle-wave, it too propagates at the speed of light when it is a wave, and therefore experiences atemporality. But, like I said, he is mistaking atemporality for the feeling of eternity; had he ever plunged all the way down through the moods of time, to the feeling of atemporality, he would have understood how wrong he was. Nothingness, including timelessness, is terrifying.

103.

If Bohr's epistemic interpretation of QED is correct (even partially) that we cannot describe nature with it, but only what we can say about nature, then Malin's position is completely undermined, since nature "in itself" may not do anything QED describes, but only shows us these aspects. Probability only expresses our ignorance of when an event will happen—'potentiality' is only an expression of our ignorance of the details of the complex dynamics of the overall system. This fortunately can open us up to an ontological interpretation by integrating what we know of QED and systems theory, rather than just declaring it all a mystery, as Malin seems wont to do. The complexity of the system does make it impossible to predict (calculate) when a wave collapse will occur, but with a mechanism, we at least know why and how it occurs. We can thus predict *that* a collapse will occur, even if we cannot predict *when* it will occur. There are kinds of predictions we need to remove from consideration in science—perhaps indeed systems theory returns science to natural philosophy, the description of nature. Not all descriptions of nature can make the same kinds of predictions. Darwin's theory of evolution cannot predict future species, but it can predict that experiments on mice will have applications to humans because of its explanation of the relationship between the two. It is only with mechanistic physics that we get the kinds of predictions in science we are used to associating with science. And even with that, we cannot prove that an object in motion will stay in motion unless acted on by another object, for we cannot have an eternity to observe such an object—we can only agree that it does indeed do this over at least a very, very long time. This claim we just accept without our ever having, or being able to have, the physical proof (though we do have the mathematical proof) that it will do so.

104.

The laws of nature are relationships and interactions, and we derive these laws by creating abstractions of those relationships and interactions—they are concepts we use to understand our observations. They are abstractions we make to explain the similarities among the different systems in the universe.

105.

Observation creates a type of actuality in QED because QED is inherently paradoxical and the paradox is resolved by observation. The emergence of each new level of reality comes about from the need to resolve the paradoxes of the level below it. Chemistry resolves the paradoxes of quantum physics in general, and of the atom in particular (the communication of information between quantum particle-waves creates observation between those quantum particle-waves). Biology resolves the paradoxes of chemistry. Human intelligence resolves the paradoxes of biology. But humans have their own paradoxes, such as the conflicts among different psychosocial levels, and the conflicts within each level that leads to the development of new ones. The next hierarchical level will resolve these paradoxes—only to create its own (J.T. Fraser). Paradox drives complexity—and makes the world more complex, more ordered, and more beautiful.

106.

Is it "the mind" or "the minding function of the brain" (as Fraser would have it)? Let us put it another way: is it the organism, or the organism-ing of the cells? Is it the cell, or the cell-ing of organic molecules? Is it molecules, or the moleculeing of atoms? Is it atoms, or the atom-ing of particle-waves? And is it particle-waves, or the particle-waving of strings? The answer, in a sense, is "yes". The danger of nouns is that people make the mistake of thinking of these things as having some sort of permanent "being." The danger of verbs is that people make the mistake of thinking that something always changing does not have any kind of form or order (dissipative structures belie this belief—but most people are not yet thinking this way). Each noun-form is the emergent structure of the action of its constituent parts—the verb-form of those elements.

107.
Each element of a system is digital in its individual existence—but it is analog in its interactions with other digital elements. A system cannot exist as a system if it does not have digital elements with analog interactions—a system cannot exist unless it is both digital and analog simultaneously. This aspect of quantum physics—its paradoxical digital- analog aspect—is what gets transmitted up into each level, to help create each level. Even if that new level is now digital and analog in very different ways. However, this aspect is not transmitted up in some transcendental fashion. Nature simply manages to come upon the same kinds of solutions over and over. When something works well, nature uses it over and over. Digital-analog systems giving emergent complexity works very well, and nature thus uses this pattern over and over. This makes sense too if the universe itself is fractal—we would expect things in the universe to be nested hierarchies of self-similar structures. Which is indeed what we find.

108.
Patterns are created at each of these hierarchical levels—and these patterns tend to be fractal. Thus, pattern-creation precedes pattern-recognition—as would be expected, since one needs patterns to recognize before you need the ability to recognize them. But why do you need to recognize these patterns? For a living organism, this is clear: everything is patterned, so if you are going to live in a world full of patterns, you need to be able to recognize them. It would also help if you could recognize when patterns are broken—this requires memory. When you can remember patterns, you can use them to make predictions about the patterns you are getting ready to see, so you can know if those patterns have been broken. There are lots of things with different kinds of memory. But when is memory used to recognize patterns and to make predictions with those patterns? If the ability to make predictions means one is intelligent, then the better one's predictive abilities, the more intelligent one is. There has to be a mechanism of comparison of input to what is stored in memory, which is part of the memory, pattern-creation system itself. Humans are most intelligent, because we have the longest temporal horizon, and are able to make the most distant predictions. Animals too can make predictions—otherwise they could neither hunt nor escape from predators. In an extremely limited sense, then, so do single-celled organisms. They have the ability to learn, so they have the ability to predict. But what about molecules? Are there molecules that can predict? Do atoms or quantum particle-waves predict? I find it unlikely (but not impossible). However, each is able to create patterns. And the creation of patterns gives the appearance of intelligence. Perhaps the best we can say is that the creation of patterns is not sufficient for animal intelligence, and that the spontaneous creation of patterns is not evidence of intelligence of this kind.

109.

The issue of patterns and our relationship to patterns is and should be important precisely because we are fractal patterned, and each system hierarchically nested within us are fractal patterned—and that our environment itself is fractal patterned. To recognize patterns of any sort—fractal or otherwise—would thus be beneficial, as would noticing that something is wrong with the familiar patterns. The more we familiarize ourselves with fractal patterns of as many kinds as possible, the better able we will be to make predictions about the world, and the healthier our relationship with the world will be as a whole. It also suggests that intelligence is immanent in the world. If the creation of patterns is a kind of intelligence, then the universe as a whole is intelligent, and has been from the beginning. If the creation of patterns is an indication that something is simply alive, and not necessarily intelligent (a better interpretation, perhaps), then the universe as a whole is alive, and has been from the beginning. And perhaps, then, intelligence in Hawkins' sense evolved as a special form of life— one that could take advantage of the fact that life and everything is patterned, and use it to perpetuate its own patterns, and to create even more complex patterns. And, in the case of humans, to create the kinds of complex patterns that could lead to even more complex levels. The living universe became the intelligent universe.

110.

The electron does and does not have a well-defined momentum. The electron does and does not have a well-defined position. The electron is a particle. The electron is a wave. All of these things are true under different conditions. All of these things are true simultaneously. Each of these is complementary. Each of these is contradictory. Contradictory? No. Together each makes up a paradox.

111.

There are no nouns—there are only verbs.

112.

There is no solid land. Our continents sail on flowing seas of magma, floating, moving, lifting, and sinking. The most solid rock is vibrating atoms with orbiting electrons. Mountains grow, valleys sink. Rivers and oceans flow. Certainly some change is slower than others—go to India and watch Mount Everest grow. Look through our strongest telescopes and watch the speed of light slow.

113.

A photon of wavelength 700nm reflects off an object and is picked up by a color receptor in the eye. The configuration change in the receptor causes an electric signal to be transmitted to the brain, which maps the signal onto a pattern associated with the pattern associated with the pattern for the word "red." Even adjectives are actions.

114.

"Chaotic motion is much more widespread than regular motion" (Manfred Schroeder, 25). If there are three or more objects in a system, you get chaotic motion. What are objects? Quantum strings are objects, of sorts. Non-chaotic motion is the special case—and it takes very special circumstances to get non-chaotic motion (if it is, technically, possible at all—if we look close enough, it may not be, as all motion may in fact be chaotic, or at least have elements that are themselves chaotic, even if the level above it is deterministically tied to the lower level). But let us add to Schroeder's statement: fractal geometry is much more widespread than Euclidean geometry. Again, Euclidean geometry appears to be the special case—and may only exist in the minds of humans and nowhere in nature itself, exterior to the human mind. When we make circular objects, we have to make them from objects in the world, and those objects have fractal borders. Euclidean math gives only an approximation to the circumference of a circular object—close measurements (say, at the Planck length of 10^{-35}m) would give a considerably different (much longer than Euclidean geometry predicts) lengths for borders of circular objects. The borders of apparently smooth-edged objects are actually rough. Rough edges are the only kinds of edges in the real world. Non-tangential curves are the only kinds of curves. Tangential curves are a mathematical fiction—but, like mechanistic physics, good enough for many things.

115.

All fractal objects obey power laws in their self-similarity regardless of scale. The smaller the scale, the more objects at that scale, in a power law relation to that scale, which is to say that the frequency is inversely proportional to some exponent of its size. "Power laws define the distribution of catastrophic events in self-organized critical systems" (Darling, David, *The Universal Book of Mathematics*, 253). Power laws are part of diaphysics—it is a rule found at all levels of reality, as far down as it goes. And as far up as it goes (and will go).

116.

The Pauli exclusion principle in quantum physics is what creates the paradoxical conditions that lead to the very possibility of molecules and chemistry (Harold J. Morowitz). It is a trimming rule, and is the kind of thing necessary for one level of reality to develop into the next, more complex, level of reality. Natural selection is another trimming rule. Natural selection through competitive exclusion creates strong, exclusionary competition, either eliminating a group, or resulting in specialization (and therefore speciation). But natural selection is not how complexity evolves; natural selection describes how complexity is selected for—in the same way that the Pauli exclusion principle is not chemistry, but what is necessary for there to be molecules.

117.

Understand how profound George Wald's statement is that the physicist is the atom's way of knowing about atoms. With the evolution of humans, "Know thyself" was finally possible for humans, life, molecules, atoms, and the universe as a whole.

118.

Can we have a view of knowledge that is neither absolutist nor relativistic? Must we have either one extreme or the other? Can there not be a probabilistic knowledge? Can we not realize that we can approach the truth through uncovering facts, but that we can never reach the truth—the absent center—the reclining noonday time?

119.

A *physis* of change is not a rejection of a form of universality or of laws of nature. There are universalities for humans—the laws of nature relevant for human conduct are unchanging in the timescale in which humans live, act, and evolved. We can know about *physis* as it is now, more or less—and use that knowledge to create a baseline to learn about how it has changed over time. Our speed is our benefit here—we can learn quickly about those things that change, relative to us, slowly. There are *a priori*s relative to us. The biological (what is necessary to even be alive) is *a priori*. The physical (what is necessary for mere existence) is *a priori*. And the diaphysical (what is necessary for organized existence leading to more complexity and order from simpler, less ordered objects) is *a priori*. If Leo Strauss is correct that "an *a priori* does not have a genesis" (*The Rebirth of Classical Political Rationalism*, 22), then there are and cannot be anything *a priori* if we take the universe as a whole as our baseline. However, relative to the human, there can be, and are, several *a priori*s, or elements that did not have a genesis *in us*. These are our "natural laws."

120.

Why science? It is a lovely tool—one which has helped us build an incredible world (and a more tragic world—one that can be more thoroughly destroyed because of science). We have both better living and better destruction through science. But creation does necessarily precede destruction. The most efficient form of prediction is through science. If Hawkins is right, and the ability to make predictions is the hallmark of intelligence, then our development of science itself made us more intelligent—why should it matter if that intelligence is, in a sense, outside of us? Should we be surprised to find that things emerge from us?

121.

Humanism is the study of what makes humans uniquely human. To do this, we first have to know in what ways we are like what came before—what is left, or how we deal with those things in a human way, is what is uniquely human. Before we can know about humans, we have to know what is not-human.

Physis

122.

The human dances precariously on the tightrope between *physis* and *nomos*. We must come to understand this. The human dances precariously on the tightrope between *physis* and *nomos*. Do you understand this? Without understanding *physis* we cannot understand what is human. We are fools speaking foolishly.

123.

Energy must act in relation to the nothing from which it comes—but humans do not come from nothing, so we therefore do not have to directly deal with it when dealing with human things. If in the beginning was nothing, and the nothing gave rise to a singularity, from which we get the big bang, and thus strings, which gave rise to quantum mechanics, which gave rise to molecules, which gave rise to life, which gave rise to vertebrates, which gave rise to social mammals, which gave rise to social apes, which gave rise to humans, we are at least eight steps away from nothing. Even if a strange attractor is an absent center—or, nothing—we still cannot ever reach it, and therefore cannot deal with it in a direct way. This is a nothing without meaninglessness.

124.

When Nietzsche attempted to overcome relativism (which comes out of historicism), he turned to nature (this is why Heidegger proclaimed Nietzsche the last metaphysician; but this was a proclamation by someone who supported one of the most tragic regimes—something that should not surprise us, for the rejection of *physis* will inevitably lead to tragedy). I side with Nietzsche. Overcoming relativism is precisely why a *physis* is necessary, and the necessity of *physis* makes ethics, politics, aesthetics, and epistemology possible. Without a *physis*, we can only have nihilism. That is why what I propose is not a radical historicism, even if it is a *physis* of change. It is tempered by a natural classicism—that which flows over and around the abyss is the ground.

125.

Mere facticity and contingency versus truth. Facts change, but we must remember that there are different rates of change. All things change, but the rules of physics change slowly, and technology changes rapidly. Consider the speed of genetic evolution versus the speed of thought. Does that mean thought and the mind (or, minding) do not come from the brain, which is coded by the genes—that there is no connection between thinking and genetics? Ridiculous! We still have genetic *a prioris* that are for all practical purposes (within a human lifespan) unchanging. Since this is the case, we have to deal with the world in this way. Thus, the need for *physis*. To put it another way: there is no conflict between cultural universals/human instincts and cultural relativism. There is no conflict between nature and environment. There is equally no conflict between the parts and the whole. Genes created organisms that created culture that created a new environment for genes to evolve in. They have, but genes are still changing slower than thought and culture. There are feedback loops between and among each of these. All are right simultaneously

126.

Aristotle's world of wisdom gave way to Machiavelli's world of knowledge. The world has taken sides: for wisdom, against knowledge; for knowledge, against wisdom (this latter is the deconstructionist view, which ironically leads one to a radically skeptical epistemology, against knowledge itself). Right and Left. No organism can survive as only right-sided or left-sided. Nor can one side of an organism fight the other—nor indeed can they be truly differentiated and differentiable. The new view: wisdom and knowledge, or Beauty. The whole organism, whole and healthy.

127.

Knowledge of the whole is indeed impossible—but knowledge of the principles (rules) of the whole (knowledge by modeling the whole) is not. Still, we'll never be certain of every detail. A missed detail can give butterfly effects—and our certainty vanishes in a wing beat.

128.

Free will and *physis*. We have free will because our mind, which is emergent from the system-interaction of the brain's neurons (and body overall), in turn influences the actions of the neurons. We see the same thing in biology: organic chemistry in system-interaction gives rise to the cell, and the cell in turn affects the kind of chemical reactions that will occur. Further, chemistry affects atomic actions (electron orbits), and atomic physics affects quantum wavicles.

129.

To have free will, we must act in an ordered world—it cannot be random nor arbitrary. The world is given order by emergence. With each new emergence into more complex order, free will became more and more possible—until it was able to emerge as a mental property.

130.

The ability to model the world (to understand it as a system), and to thus create a new kind of knowledge about the world, only came about through the invention of computers. *Techne* (technology, or physical things invented and made by man) has aided us greatly in understanding *physis* (that is, through these models, and not by false analogies—which have more often than not gotten in the way of understanding: the body is not a machine; the brain is not a computer). Without *techne*, we would have continued to have an incorrect understanding of *physis*. So this is yet another thing about which Heidegger was fundamentally wrong.

131.

If the universe expands as a Fibonacci curve, we would expect to see rapid expansion and to then have a universe that now appears both curved and flat.

Physis

132.

The way information is transmitted has changed over the history of the universe, but the fact that it was information that informed the inform to create form did not. A photon, cAMP, and language are all very different ways of transmitting information, but they are one in that they carry information from one system-object to another. Signal transmission, absorption, and interpretation is what keeps any system organized.

133.

For the creationists. — Evolution is a fact. Natural selection is a theory of evolution. Sexual selection is a theory of evolution. Punctuated equilibrium is a theory of evolution. But evolution itself is a fact. It has been proven—we are still working on the various "hows" (aside from selection of mutations—the very definition of evolution). There are no such things as "micro-" or "macro-" evolution. There is only evolution. Do mutations happen? Yes. Do bad mutations (very common) harm the organism, killing them directly or causing them to get killed? Yes. Do good mutations (very rare) cause the organism to be better? Yes. Thus do bad mutations get selected against, good mutations selected for, and change spread through the population. A change in environment plus good mutations results in evolution. Two populations of an organism in two different environments plus good mutations results in divergent evolution, and the creation of two species. These are known, well-established facts. (However, I will say that the creationists' objections—as based in error as they often are—have pointed out several shortcomings of the theories of evolution; however, these shortcomings are resolved by emergent properties theory, bios theory, self-organization theory, and information theory).

134.

Evolution in living things is not driven only by random mutations. That is but one mechanism. There are also recombinations, cross-overs, inversions—all of which are not random, but are controlled by the organisms themselves. Thus, organisms to some degree direct their own evolution—or at least direct their responses to random mutations. This is how organisms integrate other levels of complexity into their evolution—there is the random element, but also probabilistic and deterministic elements and, not surprising for living things, directed elements as well. There is thus an element of choice in evolution, within the individual organisms. And all of these are being inserted into complex, far-from-equilibrium systems, making the outcomes increasingly unpredictable.

135.

Information theoretic molecular biology: linear genes are the symbols, but proteins are the meaning, or 3-D actualities. "DNA is the word, while proteins are the deed, and deeds are more directly engaged, run more risks, confront life more irretrievably, than do words" (Campbell, 93). And we must not forget too that "there are nonsimple connections between symbols and substance in certain kinds of information systems" (94) like cells. But how seriously should we take a non-biologist's comments on biology from an information-theoretic point of view? Campbell's book was written in 1982, and yet, he suggests, based on his understanding of information theory, that in addition to coding for genes, DNA may also "consist of a kind of commentary on the structural genes, or contain hypothetical recipes for proteins which are not manufactured, or programs for editing and rearranging structural sections. They may represent information about information" (94). Indeed, every one of these things have been discovered to exist. And the science of evolutionary developmental biology has also demonstrated how a "mutation in a "nonliteral" part of the text results in "very significant modifications to the organism" (94). Experts need to listen to interdisciplinarians.

136.

Superior? Inferior? Could you do the job of an earthworm? Of an oceanic algae? Of a cow? Where would we be without each level? Our very emergence is impossible without the ecosystem—and the organisms of that ecosystem. High and low here does not mean better and worse—what is higher is held up by what is lower. We are but bricks midway up a pyramid, the base of which is made up of pure energy. And do not forget that we are but one way nature found for evolutionary success. There could have been other pathways to our level of complexity, and if something happened to the human race, there could be other pathways back to it. And don't forget too that, on the Earth at least, it took 3.5 billion years to get around to us.

137.

50% of the human genome is shared with yeast.

138.

Natural selection and sexual selection do not result in increased complexity. Emergence from complex systems and nonlinear dynamics, however, does. Natural and sexual selection are both ways that these new emergent, complex aspects of the new organisms are selected for or against. Punctuated equilibrium and biotic processes appear to be the best descriptions of how new complex organs emerge in organisms. But once complexity emerges, the first two likely take over—at least for a while.

139.

On causality — What is *the* cause? *The* cause? The last time there was a single cause of anything, the singularity caused the big bang that resulted in the universe. Since then, everything has had multiple causes—and with each emergent level, the causes have multiplied. And in quantum physics, causality becomes questionable at best.

140.

Rules come about through individual elements interacting with each other as they self-organize. The best rules create unbounded, generative, creative systems. Generative rules allow for unbounded systems, which contain "more information than we might expect by merely looking at its component parts" (Campbell, 128).

141.

What is a fractal? It has fractional dimensions, and a finite space encompassed by an infinite border. But how is it infinite? Because all fractals exist in time. If you stop the iteration process, you will, of course, have a finite, measurable border. But if you measure the border as the iterations continue, the border will reach infinity. As time approaches infinity, so does the length of the border. Yet the area remains finite.

142.

The universe is a fractal. Every atom, molecule, object, living thing, and human is part of the infinite border of the universe. In between is nothing. All the rest is spacetime.

143.

Over time, the universe has become more complex. Humans are more complex than any other organism. Any organism is more complex than any nonliving system. Any nonliving system is more complex than its constituent molecules and atoms. Over time, as the universe has become more complex, its border has become more complex and, thus, more and more approaches infinity in length.

144.

From the reductionist standpoint, physical law is the motivating impulse of the universe. It does not come from anywhere and implies everything. From the emergentist perspective, physical law is a rule of collective behavior, it is a consequence of more primitive rules of behavior underneath (although it need not have been), and it gives one predictive power over a limited range of circumstances. Outside this range, it becomes irrelevant, supplanted by other rules that are either its children or its parents or in a hierarchy of descent. (Robert B. Laughlin, *A Different Universe*, 80)

This is related to "Goldstone's theorem, the statement that particles necessarily emerge in any matter exhibiting spontaneous broken symmetry" (107). In other words, "The myth of collective behavior following from the law is, as a practical matter, exactly backwards. Law instead follows from collective behavior, as do things that flow from it, such as logic and mathematics" (209). Thus, too, human law and human society, economy, and culture.

145.

New levels of complex behavior emerge from the collective behavior of elements of a lower level. Beehive behavior emerges from the work of each individual bee, and not from some master bee coordinating behavior. Economic behavior emerges from the actions of each individual human being, not from top-down regulations (and often in spite of them, as Henry David Thoreau observed).

> Emergence means complex organizational structure growing out of simple rules. Emergence means stable inevitability in the way certain things are. Emergence means unpredictability, in the sense of small events causing great and qualitative changes in larger ones. Emergence means the fundamental impossibility of control. Emergence is a law of nature to which humans are subservient. (Laughlin, 200-1)

In fact, "human behavior resembles nature because it is part of nature and ruled by the same laws as everything else," and "The parallels between organization of life and organization of electrons is not an accident or a delusion, but physics" (201). Or, diaphysics.

146.

When "observed behavior remains the same if the sample size is doubled, followed by a corresponding doubling of the scales for measurement of distance and time," then we have a process called *renormalization* (Laughlin, 146). In other words, renormalization is the process of having self-similarity regardless of scale. Where does this self-similarity regardless of scale come from in the universe? Because "The vacuum of space is renormalizable" it "is known to be near phase transitions. There are numerous experimental indications that the vacuum emerges in a hierarchy of phase transitions in which the various forces of nature differentiate from each other" (154). The scale invariance of space itself means that space itself is fractal. Further, if space is at phase transition, then space is a dynamic system with fractal geometry—including self-similarity regardless of scale that will inevitably give rise to emergent complexity, as this is precisely what happens at phase transitions. And since space and time are not separable, time too is fractal and emergent.

147.

There are *physis*-laws and *nomos*-laws—laws that give rise to entities, and laws that evolve from entities and, in turn, affect the actions of those entities. The *nomos*-laws of a lower level are the *physis*-laws for a higher level. The *nomos*-laws are the laws of the level emergent from the lower level, the entities that, working together as a system, give rise to the new, emergent level.

148.

Neoteny—the retention of youthful traits—has resulted in every one of the steps in emergence of more complexity in the universe, at each level: physics, chemistry, biology, and human, as well as into each level.

149.

Humans are neotenous apes. Vertebrates are neotenous sea squirts. Multicellular organisms are neotenous polycellular organisms (they have the qualities of both polycellular organisms and single-celled organisms). Eukaryotic cells are neotenous combinations of primitive archaebacteria, which gave rise to nuclear DNA, and eubacteria, which gave rise to the mitochondria and chloroplasts. Cells themselves are neotenous between chemistry and quantum physics in, for example the way proteins fold (with quantum tunneling, according to Satinover). And complex systems chemistry (the kind which gave rise to life), is neotenous between solid-state chemistry and fluid dynamics. Stars are neotenous insofar as they create more complex atoms (the "adult" stage of quantum physics) through making use of the free particle-wave state of quantum physics (its "immature" stage). Thus, more complex atoms are created through a neotenous process—which establishes neoteny as one of the creative processes of the universe.

150.

With the emergence of each of the *umwelts* of reality (to return to J.T. Fraser's model), each new level repeats, at its level, in a self-similar fashion, each of the levels that preceded it. The first level has only one sublevel—atemporality only contains energy.

151.

Prototemporality has two levels. The first one is that of free particle-waves. The second is the more complex, more "solid" atoms, which can engage in more complex chemistry.

152.

Eotemporality has three levels. One is fluid dynamics. The second is solid-state physics and chemistry, which give rise to mechanistic physics. The third is complex systems, including complex systems chemistry. To some extent, the third is driven by aspects of the prototemporal level (making it neotenous in another way).

153.

Biotemporality has four levels. The first level is that of single cells, which tend to act in similar ways, regardless of whether they are eukaryotes or prokaryotes. They are a level above complex systems chemistry. The next level is that of multicellular organisms, which include plants, fungi, and animals. With the animals, we have the line that led to vertebrates, invertebrates in the line that led to those with exoskeletons, and the predecessors of both. And in the vertebrates, we have schooling/herding, territorial independent, and territorial social animals (the latter appearing to be a neotenous combination of the other two).

154.

Nootemporality has five levels. Don Beck recognizes six levels of human psychosocial development, but his first one actually belongs as much to other social apes, especially chimpanzees and bonobos, as to humans. Beyond that level, which belongs in the biotemporal level, there are five distinctly human levels of emergently complex thinking. The first is tribal/familial thinking. The second is the egocentric level, when writing is invented in a culture, during which we hear tales of people typical of this level, heroes such as Achilles, Odysseus, and Gilgamesh. The third is the level of truth and authority, typified by such thinking as that of the medieval Christian church and contemporary social conservatives. The fourth is the level of capitalist and deterministic scientific thinking, typified by materialism and liberal/libertarian thinking. The fifth is the level that promotes human bonds, and typically is anti-hierarchical, relativistic, perspectivist, egalitarian, and often respectful of differences among people. It seems strange to suggest that this is the last of the human ways of thinking, but I have come to the conclusion that Don Beck's Second Tier thinking, which is only just beginning, is a new emergent level of reality.

155.

J.T. Fraser says that the next *umwelt* is the sociotemporal level, which he says emerges from the interaction of humans in society. But this makes as much sense as saying that the level emergent from the biotemporal is the world ecology level of time experience. The nootemporal, the actual next level of complexity, evolved from a particular kind of animal, and the next level of complexity will evolve in humans. In fact, I would argue that it already has evolved, and it is Beck's Second Tier thinkers—which will have six levels proper. The first two are the integrationist level, and the holistic level. Others will follow and, if the same pattern that has been found in the universe throughout its entire history holds, it will have six of these sublevels before there is a movement into the next level. And, just like humans have all the other less complex levels in them (the biotemporal, etc.), the Second Tier thinkers are indistinguishable from other humans, except in their thinking. I propose calling this new level not sociotemporal, but intertemporal, since it is the first level that is self-aware both in understanding the depth of its thinking, through all the sublevels of the nootemporal, and in understanding the nested hierarchical temporality of the universe as a whole. In this sense, the intertemporal level is also neotenous, in that it takes on all the levels of human thinking, embracing them all, but also every other level of reality, and not just nootemporality.

156.

There is a correlation between flattened fitness landscapes and leaps into new levels of complexity. The egalitarian level is an example of a flattened mental fitness landscape, thus preparing the way for the emergence of the next level of mental complexity. All hierarchies are eliminated in such thinking, preparing the way for new hierarchies, at the new level of complexity. Humans live everywhere in the world because we have a flat biological fitness landscape. The same is true with vertebrates. It is no coincidence that both arose through neotenous processes.

157.

Determinism is unidirectional. But what happens when an emergent level then affects the level(s) below it? Not only does the simpler system determine the more complex one that emerges from it, but the more complex one determines the (inter)actions of the simpler systems that constitute it. Iit is bidirectional, a feedback loop that makes the entire complex system nondeterministic and inherently unpredictable. This is true between every level of complexity, as each higher level is made up of a nested hierarchy of each of the lower levels—meaning the more complex something is, the more complex, and less deterministic, the behavior. Quantum physics affects chemistry, but chemistry in turn affects the kinds of quantum effects that occur. Chemistry creates cells, but cells in turn affect the kind of chemistry that occurs. Biological processes make human intelligence possible, but human intelligence in turn affects our biology through things ranging from belief to medicine. And if I am correct that human intelligence has given rise to integrationist intelligence, then integrationist intelligence will in turn affect the way people think (and the way those people communicate with each other, since there is an emergence into new kinds of communication with each emergence into new levels of complexity—such as the emergence of grammatical language in humans), and will then drive the evolution of integrationist thinking in a nonlinear, nondeterministic feedback loop. But we will accept that this is true only if we accept that the universe will continue to use the same processes it has always used throughout its 15 billion years of existence.

158.

All bonds are created through information transmission. Human bonds are created through language. The bonds of social mammals are created through touch and grooming and transmission of information regarding dominance and submission, which transmit information regarding social status and closeness within the group. The bonds of cells are created through transfer of molecular information. Chemical bonds are created through transfer of electrons. Atomic nuclei are bonded together through the transmission of gluons, which, in being passed back and forth, pull the elements of the nucleus together, despite the fact that any nucleus larger than a hydrogen nucleus has multiple protons, and thus have charge repulsion. But isn't it this attraction-repulsion that creates more complex atoms, that create the bonds within the atom? The same is true of all bonds—atomic, chemical, biological, social, human, cultural, love, or economic. Attraction and repulsion are found at every level of complexity in the universe—the bonds they create are different, but self-similar, and created in self-similar ways. The kinds of attraction and repulsion differ, but the fact that they are attraction and repulsion do not.

159.

Living things evolve on what are called fitness landscapes. A fitness landscape is a mathematically idealized topology that describes the fitness of organisms according to whether they are on a hill, a mountain peak, or a valley. The higher the peaks, the higher the fitness. But there can be local peaks (hills) on which an organism is stuck. Sometimes extreme fitness in one environment means extreme maladaptation for another environment. So if we have a highly fit species in one environment, and the environment changes too much (and the environment always changes), then the species can and likely will go extinct. The mountain collapses, and the organism is now in a valley. However, "The best exploration of an evolutionary space occurs at a kind of phase transition between order and disorder, when populations begin to melt off the local peaks they have become fixated on and flow along ridges toward distant regions of higher fitness" (Stuart Kauffman, *At Home in the Universe*, 27). Mentally, the human species is always in the edge-of-chaos realm, which keeps their fitness landscapes melted, and thus smoothed out, so that they are able to adapt to any environment.

160.

By having a melted fitness landscape, humans are able to have a sort of biological order. The human mind allows us to adapt to any climate through the use of technology. With the emergence of more complex ways of thinking, the emergence of a more chaotic mental realm on the edge of chaos, the human brain keeps the human fitness landscape melted. Absent the abilities our brains give us, we would only be able to live in tropical regions, like Africa. But our brains keep our fitness landscapes melting, so we can live literally anywhere on earth. All physical adaptations to environment are tiny at best, and easily reversed through outbreeding with other human populations, which are genetically practically identical to every other human population. It is likely in no small part due to our brains creating such complex levels of thinking, and thus such complex cultures, that we have such extreme genetic order, as we are practically clones of each other.

161.

Every fitness landscape for every organism does not exist in isolation, but is coupled with every fitness landscape of every other organism in its environment. As the fitness landscapes for any organism or species changes, so does the landscape for all the other species. The more orderly a species (that is, the more specialized it is), the more determined it is by what happens in its environment. An orchid that is only pollinated by one moth will go extinct if the moth does. But if a species is more adaptable—a generalist rather than a specialist—then changes in their environment have less of an effect on them. To emerge into more complexity, specialists have to first become generalists—in physics, chemistry, biology, or even human thought.

162.

Art and philosophy take over explaining the world at the incalculable.

163.

Most things are incalculable. Human things, especially.

164.

Science is a useful way of understanding the levels of complexity below the human. But only the humanities can be used to help us understand the levels of complexity emergent from the human. Both are needed if we want to understand the world as a whole, in all its complexity, and all its levels of complexity.

Physis

165.

In a complex system, context is everything.

166.

What is matter? Here's a tautology: matter is what materialists think the world is made of. Matter takes up space—but this is only true if matter and space are two separate things. We also make the same separation between matter and energy. But all of these separations should not be made. Matter is energy is spacetime is matter. The physicality of matter evolves from the nonphysicality of energy, which are both different complexities of spacetime folding. It is undifferentiated, formless, able to evolve, informed, substantial, and form. Matter is action, quantity, amount, extent, set, set down, inscribed, and of importance. Matter matters. But only once it is made to matter.

167.

First tier thinkers are the universe contemplating its own existence. Second tier thinkers are the universe aware that it is contemplating its own existence. Reflexivity is what gives rise to leaps into new levels of complexity. Animals are aware of themselves, but are not aware of that awareness. Humans are aware that we are self-aware, and thus are able to contemplate the universe — the universe becomes self-aware for the first time. But with the next level, the universe becomes aware of itself being aware. Thus the universe gains in intelligence, information, consciousness, complexity and self-awareness.

168.

We talk about how observations affect whether we see an electron, for example, as a particle or a wave, but we forget that all things observe all things, and thus affect our behavior. We observe the electron, but the electron observes us too. The difference is merely at what level of complexity each is observing the other. But both transmit information — each to the other — and that is how each observes the other into particular behaviors. A person who sees me as a friend observes me into friend-behavior. A person who sees me as a hotel front desk clerk observes me into hotel front desk clerk behavior.

169.

When we observe an electron into behaving as either a particle or a wave, we are asking it a question, and it answers us. "Would you mind acting like a wave, so I can observe your behavior?" It answers by acting like a wave. What else could we ask it? How else could it answer us?

170.

An alternative title for this work: ecophysics. If we should have balance between *physis* and *nomos*—and we do so through *logos*—then since we have *oikos-nomos* and *oikos-logos*, we should have an *oikos-physis*, to ensure we are getting the former two right. Ecophysics—Ecology—Economics. We need to get our house(s) in order. We need to at least recognize that a house need a foundation.

171.

Bios theory shows that growing, changing, creative systems are nonlinear and exist on the borderlands between order and disorder. Complete order—and complete disorder—both are definitions of being dead. A salt crystal is an example of complete order—a gas in a closed container at a constant pressure and temperature is an example of complete disorder. Living things exist on the edge of order and disorder, the realms of chaos and bios, wherein lies the principle of growth. Living things are systems, and systems have order and disorder—the heart is a system (which part of the circulatory system, which is part of the organismal system) that can have neither completely orderly beats, nor completely random beats, but must have beats that are mostly orderly, with some disorder, which means the beats are biotic, creating a mandala shape when their time series is transformed by sine and cosine, complementary opposites. Cell membranes are orderly and disorderly—they are liquid crystals, fluid yet solid, as the proteins and phospholipids slip past each other. All living things live on the principle of growth, live in the realm of order and disorder, live lives far from equilibrium.

172.

If the principle of growth and stability for life is in the nonlinear, far-from-equilibrium realm between order and disorder, this would also be the principle of growth and stability for systems of living things as well, including superorganismal systems such as ecosystems, economies, and governments. Indeed, studies of ecosystems show they are not stable—at some sort of equilibrium—but are in fact always in flux, always changing, in time. And the way they change follow power laws—with many small changes, a few medium-sized changes, and very few large changes, as we see in avalanches of sand when we pile sand up one grain at a time. They are systems far-from-equilibrium, always growing, in a state of orderly disorder—disorderly order. If something were to happen to make any given ecosystem stop changing—which is to say, stop growing—that ecosystem would die off. Ecosystems are stable only so long as they are constantly in flux, constantly changing. Thus, they cease being the ecosystem they are within the next moment, forever changing—deserts move in and recede, forests expand and recede, grasslands expand and recede. Meandering rivers cut off oxbow lakes where new kinds of fish evolve—to be introduced to the river when the meandering river merges again with the oxbow lake. The new fish compete with the other fish in the river, pushing some to evolve, others to go extinct, others into other habitats. They change as the river changes, flowing into new species with the flow of time and the flow of the very river in which they live.

173.

Nothing. Unstable nothing gave rise to the singularity. Energy burst forth, the Universe was born in spacetime. Energy rolled back and forth across spacetime, in solitons that pushed spacetime out, causing the universe to expand. These solitons, waves, pushed out and interacted with each other, creating more complex patterns of waves. Waves interact with waves to create more complex patterns of waves, trillions of wavelets that interacted with each other in simple systems to create the first particle-waves of matter-energy. Matter came about through the folding of spacetime, and the interactions of those folds to create the first, simple, complex systems. And when these folded back onto themselves, creating folded folds, the first atoms emerged. And these new folded realities folded again, and molecules were born, and this folded once again, and living things were born into the world. Complexity emerged with ever more folds, and as living things enfolded themselves, more complex organisms, including vertebrates, were born. And more complex vertebrates emerged from more enfolding, and more complex nervous systems emerged with more neural enfolding, creating more complex behaviors, driving even deeper, more complex neural folding. And human intelligence emerged, in deeply folded neurons, in a deeply folded brain. And then . . . ? What, indeed, is the next "and then . . . ?"

174.

What have I in fact just said? I have said, first, that all existence is nothing more than folded spacetime. It is ever-enfolded and enfolding spacetime. Any entity is not in spacetime, it is spacetime. It constitutes and is constituted of spacetime. If gravity is curved spacetime, then the curving is caused by the pull created by the folds—by the external folds, those folds that are "external" on the system. In a less densely folded object, there is in fact more externality to the system, since the folds are farther apart. A sun is both heavier, and less dense than the planet Earth—it thus has less dense folds, though many more, though simple, objects in it. This is what bends spacetime. The earth is denser in that it has more complex, and thus more densely folded, atoms in it—and this is what causes it to have the gravitational pull it has. One may suppose, then, that if this model is correct, molecules should have greater density than atoms, and thus must weigh more. This is obviously incorrect. But one thing we have failed to take into account to this point is the fractal geometry of the folding. When we have a fractal object, we have an object that has a finite area surrounded by an infinite border. With a real-world fractal, we have a finite area that can be surrounded by an extremely and increasingly long border. The addition of more folds creates more border, but the amount of space occupied remains the same. This is what allows for a relatively moderately-sized bipedal ape like ourselves to nonetheless have very complex brains—the folding encompasses a similar area while increasing the surface area. Certainly, brain size increased as humans evolved, and this helped with the creation of even more complexity—the same way that a uranium atom is both more complex and larger than a hydrogen atom—but we can compare the human brain to an animal with a similarly-sized brain, and the human is still more intelligent, and has a more complex brain. So the fractal model still holds. The human brain in fact has many more folds in it than does the brain of any other mammal with a brain of similar size, and yet takes up the same area. It also has six rather than three surface layers. More, human brains take up much less area than does an elephant brain, or a blue whale's brain, and yet humans are much more intelligent, due precisely to the deeper folds and extra layers, creating more surface area.

175.

But if greater complexity is caused by more folds in spacetime, and all folds in spacetime are in fact energy waves, doesn't this mean that there is in fact more matter in a more complex entity? After all, we all know that $E=mc^2$. Well, in a sense, there is more mass in more complex entities—just not in the same way as exists in relatively simple objects like atoms. Another thing to consider is that when we split an atom, we are reducing the atom, which is one of the lowest levels of folding, to the less folded pure spacetime. It is a violent reaction, but it is in fact a low-level one. Higher levels of folding must first unfold into lower levels, and go down level by level. The human can be reduced to the animal, an animal reduced to molecules, and molecules reduced to isolated atoms. Thus does the energy level of an entity decrease, level by level, the way that high-energy electrons drop down electron shell by electron shell. In this way, the matter of more complex entities increases, while the matter stays, in one sense, the same. What we will need will be new ways of measuring "mass" and "matter" in order to understand how deeper folds of spacetime give rise to more complex entities.

176.

One way of looking at things is to see spacetime itself as the real, and emergence into new levels of complexity as new levels of quasi-reality. Thus, it is spacetime that is the real, while particle-waves are quasi-real, atoms are quasi-quasi-real, molecules are quasi-quasi-quasi-real, etc. through biology and humans. Thus, in a sense, each emergent level is more "ideal," and the philosophies of "realism" and "idealism" are allowed to co-exist along a continuum. Another way of looking at things is to see that each time more spacetime is folded and enfolded, more spacetime comes into contact with more spacetime. Thus, there is emergence into ever-more complex levels of reality. More spacetime is experienced by a complex entity, and thus it becomes more real. With this latter view, we are able to see how nothing can give rise to something, and not only that, see the way in which that something grows, and grows more complex.

177.

Power law distributions result in self-organization. And self-organization is done according to power laws. It is a complex feedback loop. The universe is self-organizing. Therefore, it is organized according to power laws. In a complex feedback loop.

178.

With nonlinear dynamic processes, we get discontinuities, or emergence. This is true of organic molecule systems, biological systems, mental systems, social systems, or economic systems. Not even learning is purely, or perhaps even particularly, gradual. Throw out your linear models of learning—they are wrong. Brains are not linear systems.

179.

Bifurcation mechanisms reduce entropy or noise by separating elements into relatively stable attractors. This can allow for a "success" and a "failure" attractor, resulting in the emergence of many more forms, as many can now be tried out. This can also allow for splitting off into new species—of elements or organisms. But even failures need not remain so. Jiggling a system—and in a dynamic universe, systems are constantly jiggled—can turn failure into success as the system leaps into a different attractor.

180.

More information added to a system can melt "walls" between attractors, combining, rearranging, evolving systems.

181.

"*Self-organization* is essentially a *learning* process" (Stamovlasis, *NDPLS*, 10(1), 61). The system learns to become what it is.

182.

Some theories of information communication and creation —. Evolutionary theory describes how information changes and interacts to create change. Chaos theory, bios theory, and fractal geometry describe the structure of information as both ordered and disordered simultaneously, and as self-similar regardless of scale. Complexity theory is really part of systems theory, and systems theory describes how information communicates to create complex interactions. Self-organization theory describes how new kinds of information are created. Power laws describe the organizing principle of self-organization. Bios describes how new kinds of information are created. Cybernetics describes how order is maintained in a system — it is information on how a system keeps and maintains its form once informed. "Cybernetics enforces consistency. It permits change, but the change must be orderly and abide by the rules. It is a universal principle of control, and can be applied to all kinds of organization" (Jeremy Campbell, *Grammatical Man*, 23). All of these work, separately and together, to create more and more kinds of information—meaning, more order and, therefore, more complexity—in the world.

183.

$E=mc^2$ because E becomes m through information. Mass is informed energy.

184.

To say "folded spacetime" is to say "solitons of space" or "waves of space". All I've done here is move the time reference from the noun to the adjective, turning the adjective itself into a noun. Does the addition of time to qualities turn those qualities into objects?

185.

"Vacuum" is a quality of space. A vacuum is a volume of space empty of matter. Space has the near-density and pressure of a vacuum. It is near, because each particle is a probability function in space, so space is never really a vacuum; it is actually full of particle-wave potential. Another way vacuum is defined is with quantum field theory and string theory, where a vacuum is the ground state in the Hilbert space; that is, it is the state with the lowest possible energy. But, again, this remains a quality of space. Space is what is fundamental. Spacetime is a geometrical model that describes the geometry of space and time. What I am proposing is a geometrical model. This most fundamental of quantities—space—is curved. There are various ways in which it is curved. When spacetime is curved in the presence of a massive object (a massive object here is an object that has mass; it is not "massive" in the more common use of the word, but rather in the physicists' use of the word), spacetime curves. This space curves not just in the presence of massive objects, but as part of its very nature. Space itself is inherently curved. When we have a curve in motion, we have a soliton—a wave. These space-waves can have very long and very short wavelengths. And they all interact with one another. This results in their informing each other, creating the zillion tiny wavelets that result in the emergence of the particle-forms of quantum particle-waves. Quantum information theory supports this idea, since "Even a small number of qubits allow an extraordinarily rich texture of interfering waves as they compute" (Seth Lloyd, *Programming the Universe*, 138). Thus, ontologically, everything is waves of space, and these waves of space are informational -- they have no form, and yet are capable of giving form and of taking on form. When we ask "waves of what?" we are continuing to think at the intermediate level of physics -- where we have waves of water, with energy moving through the water. But space is not a "what" in the same way. The waves move through nothing. Thus, what is fundamental is the wave itself. Since it moves through the vacuum, it is inform.

186.

In a sense, since gravity is understood as curved spacetime, it is gravity which is fundamental since, if gravity = curved spacetime, and folds are curves, and everything is made of folded spacetime, then everything is gravity. The more densely packed together these waves are, the more effects (which are always curvature of some sort) they have on the local spacetime geometry, which we interpret as gravitation. Other interactions give rise to other kinds of waves—and all these waves communicate information to other aggregates of waves.

187.

Objects as large as planets are attracted to each other due to the curvature of spacetime. They move toward each other and interact to form a system. Information is that which is inform which gives form. Each planet has an inform gravitational field that allows the planets to interact to form an orbital system. The curved spacetime in a real sense "sends word" that there is something massive in the area. That is precisely what curved spacetime does: as something comes into the gravitational field of the massive object, it begins to interact with that object. It can only do so because each are in the area, and each have gravitationally interacted. If we are talking about curved spacetime, that "word" is the altered geometry. If we are talking about gravitons, that "word" is a graviton.

188.

I speak of systems, because everything is made of systems. The universe as a whole is a system of interacting waves. The brain is a system. Organisms are systems. Cells are systems. Atoms are systems. And the fundamental elements are wave systems. If we do not understand systems, we cannot understand anything. Information communication among the parts is what holds systems together and allow them to be systems.

189.

If much of this appears to be a poetic understanding of things, please note that, like poetry, physics is only accepted when it is shown to be "elegant" — or, beautiful. Physics is not so far away from poetry as we think. And poetic understandings—metaphorical understandings—are all we can only ever have anyway. This does nothing to devalue mathematical precision (math must also be elegant to be accepted). People like Rothko have shown the beauty of minimalism. But we must not forget poetry—for it is through poetic understandings that we come to truly understand what is happening in the world. Sometimes you have to reduce things to the simplicity of mathematics to understand things; sometimes you have to rise to the complexity of poetry to do so. All the great physicists have been true poets.

190.

We can understand the slowing of time in strong gravitational fields and high velocities, and include J.T. Fraser's idea of emergent levels experiencing more time, if we understand space as folded — or spacetime as a wave. A spacetime wave is merely a moving fold of space. We can imagine time as the movement of spacetime waves. Further, if we imagine the now of time the crest of these waves, we can begin to make sense of many of the issues raised when considering problems of time. According to quantum mechanics, the smallest unit of time is 10^{-43} seconds — Planck time. Equally, the Planck length is 10^{-43} m. So it seems that the smallest wave of space has a wavelength of 10^{-43} m. This would also be the rate of transmission, every 10^{-43} s. But not all waves are of this size or move at this speed. These are merely the smallest and fastest. When we have an object with mass, it has two qualities. One, it is very densely folded. Two, it bends space toward it. Time is experienced as slower in both cases, because there is more space for Planck-length waves to traverse. At higher velocities, an object is made more dense (this is what happens when it shortens), meaning it contains more folded space, meaning time appears to move more slowly. It actually moves at the same rate — there is just more space to move through, even though it does not appear to be the case. As a consequence, due to human spacetime density (or, better, "human wave density"), our "now" is, according to Frederick Turner and Ernst Poppel, 3 seconds long — the length of our short-term memory (*Natural Classicism*). A human is made of waves of waves of waves of waves of waves of waves. Due to this density, what once took 10^{-43} s. now takes 3 s. But this also means our "now" is much more densely packed than a Planck "now." This is how each emergent level experiences more time.

191.

The great thing about waves is that they move due to energy (which takes care of the 1st law of thermodynamics) and they form patterns when they interfere with one another. The more complex the interactions, the more complex the forms made. However, the more complex the forms, the less we can know about them. In a real sense, "entropy, which is just invisible information, is also a measure of ignorance" (Lloyd, 80). So the 2nd Law of Thermodynamics is about information.

> The law states that entropy (which is a measure of information) tends to increase. More precisely, it states that each physical system contains a certain number of bits of information, both invisible information (or entropy) and visible information—and that the physical dynamics that process and transform that information never decrease that total number of bits. (66)

Further, the 2nd Law "holds that the total amount of information never decreases" (76).

192.

The 1st Law is about the conservation of energy. The 2nd Law is about the conservation of information. Another perspective: the 1st Law is a theory of movement, since "Energy limits speed" (Lloyd, 158) and the more energy there is in a wave, the more speed it has—meaning the 1st Law is about the conservation of movement. We can thus see the relationship between energy and time. The 2nd Law then is a theory of interaction. Some of that is visible, some invisible. The tendency is for it to become increasingly invisible (unknowable). But this also allows us to understand too how the 2nd Law allows for the creation of greater complexity in the universe.

193.

The universe tries to remain at symmetry—since the big bang, that symmetry has been a symmetry between equilibrium and far-from-equilibrium states, the latter of which has the quality of symmetry-breaking. When a system is at equilibrium, it is not symmetrical—thus, the system bifurcates to create symmetry. But far-from-equilibrium states cause bifurcation, and the bifurcations proliferate as the set of systems created by the far-from-equilibrium state, until a new equilibrium is discovered.

194.

The evolution of vertebrates, resulted in a search for a new genetic equilibrium for vertebrates. The vertebrate structure provided a stable structure that allowed for a level of experimentation in body form that resulted in the proliferation of forms, from fishes through amphibians, reptiles, birds, and mammals. Each was part of the search for genetic equilibrium, stability of form. In one branch of apes, the hominids, this state was reached. *Homo erectus* was able to spread throughout Africa, Europe, and Asia precisely because it was highly adaptive and, thus, at genetic equilibrium. Population dynamics took over natural selection. However, this does not mean that mutations did not occur within the population. Those that did occur spread rapidly throughout the population. When a population becomes isolated, it is in the position to be affected by natural selection—and this seemed to have occurred in a population of *Homo ergaster* that first became *Homo heidelbergensis*, and then *Homo sapiens*. This species had the advantage of being both in equilibrium—at the genetic level—and of being in a far-from-equilibrium state—at the mental level. The genetic and structural stability of the vertebrate form allowed for the creation of a new level of instability in the way the brain functioned. With the emergence of language, the human mind was born, and with it the proliferation of cultural variety we are now familiar with.

195.

The output of the human brain is truly staggering. Think of all the languages humans have ever spoken, of all of the religions and myths in which humans have ever believed, of all of the technology humans have invented. This is the result of our minds being at a far-from-equilibrium state. Within the rather strict vertebrate form (including the hindbrain, the midbrain, and the mammalian cerebral cortex), and within the remarkable genetic stability of humans, the fourth (distinctly human) brain is highly plastic and, thus, highly generative of new things—whether those new things include new religions, myths, arts, technologies, etc. The products of the human brain are similar to the products of vertebrate evolution—new forms evolve and evolved, some of them survived, some of them became extinct (and fossilized, so we could study them later). Those that survived mutated and/or hybridized (and then mutated) to become new species, with new possibilities. And just as in eukaryotes, where much of the junk DNA is a record of the past, and can act as a source of future innovation, the remnants of the human past, of human past achievements, some of which have lain dormant for centuries, remain available for future innovation—wherein the past is available to create the future. In fact, mining the past for the future is precisely how one gets jumps in innovation and complexity. The most radical mining of the past—the retention of youthful traits into adulthood, or neoteny—seems to have been the primary driver of leaps into new levels of complexity, including the leap from being mere clever mammals to full humanity. All our cultural creations are our cultural environment, and thus part of the human mind, only externalized. They create more complex systems, and contribute to the increasing complexity of human thinking, as developed by Clare Graves, Don Beck, and Christopher Cowan, and undoubtedly have contributed to the emergence of second tier thinking—a new, more complex level of thinking beyond the human, but which includes the human.

196.

Every product of the human mind has the potential to be a strange attractor. Every cell type in a given animal is a strange attractor, just as every species is and was a strange attractor. With cells, the basis is the DNA—identical DNA—that is used in various ways to create the different cell types in a given body. In vertebrates, it is the central nervous system and internal skeletal structure, used in various ways to create the various species of vertebrates. The cheetah body form is a strange attractor—just as is the ant body form, the ostrich body form, and the human body form, for that matter. Cheetah DNA creates a trajectory that creates the cheetah body form. In the same way, the human brain biotically produces ideas that can create a trajectory that attracts other human brains to create a system. The major religions—Christianity, Islam, Judaism, Buddhism, and Hinduism—are all very large strange attractors, so large in fact that they have spun off smaller attractors within the larger system. The superattractor of monotheism created Christianity, Islam, and, the parent of both, Judaism. Out of Christianity we have gotten the smaller attractors of Catholicism, Orthodoxy, and, from Catholicism, Protestantism, which itself spun off Baptists, Methodists, Anglicans, Calvinists, and Lutherans, among others. And the current Baptists, which evolved from the Anabaptists, have split off into Independent Baptists, Missionary Baptists, and Southern Baptists. Mutations, fusions, and bifurcations have created an ever-increasing number of kinds of Christians. The same is true of any sufficiently large religion. Anyone familiar with the New Testament of the Bible knows that even the Jews were, over two thousand years ago, split into Pharisees and Sadducees.

197.

Every product of the human mind that attracts others to it in such a way as to create a kind of community is a strange attractor—and every product of the human brain is a result of mutation, recombination, and bifurcation. It was two groups of bicycle repairmen who invented both the car and the airplane. The bicycle bifurcated and mutated, in part through recombination, into automobiles and airplanes. How much of our lives now orbit around these modes of transportation? We make sure our cars are taken care of, that they are kept healthy and well-fed (wars have been fought to make sure they are well-fed). We build roads for them. We house them, clean them—some people even name them. We come to learn each vehicle's "personalities," we have favorite cars, talk about them as if they were living organisms, and sometimes even mourn the loss of a favorite car. We work long hours to be able to afford to buy cars—others work long hours to make cars. Do we control the cars, or do the cars control us? The invention of the car caused an intricate system to form around them. And we, the creator of the car, are the main components of that system.

198.

But to say that ideas emerge from us and create systems is not to say that every system is as good as every other system. Species go extinct when they become maladapted. We no longer have trilobites or sauropods. Other species that seem to no longer be "needed" continue to find their niche—the lancelets that acted as the bridge from sea squirts to vertebrates continue to exist within their niche, despite the fact that other species have come along that are better adapted. The lancelets continue to survive mostly because they stay out of the way of everything else. The same is true of various ideas. Astrology continues to exist despite the fact that there is almost certainly nothing to it—it has outlived its usefulness, but at the same time, it doesn't seem to harm anybody who does believe in it, so it continues to survive. Other ideas, like Marxism, have proven to be extremely harmful and destructive of the very components of the systems created around them, and so are now mostly gone or on the way out. The important thing is that an idea creates an even more complex system around it—the creation of more complex systems is a measure of the value of that idea. The car is a good idea, because the system created around it is highly complex and generative of greater complexity (this does not mean that everything that has come about because of the car is necessarily good—just that the system created is, as a whole, good). Marxism is a bad idea, because the system created around it eliminates complexity and is highly destructive of others and, even, itself. It is based on a false understanding of the world. The question is always if the idea giving rise to the system is creative or destructive. A cell, for example, never destroys the atoms that make it up—there is only ever transformation and/or (if the molecule made is toxic) expulsion. And any cell that makes more toxins than good molecules will eventually, and rightfully, die. The best system is in fact that system which generates the most new things fastest without poisoning itself (or others). If this is a proper measure of what is good—the good being that which makes the universe more like itself—then the creation of more, and more complex, things is good, while stagnation and creationless destruction are both bad. Thus, as humans continue to create more and more things, and more and more complex things, so long as those things are not toxic to their environment-system, we become increasingly good. We are good when we value the kind of diversity that excludes and expels toxins—those things that are destructive of diversity, life, and creativity.

199.

Certain ideas, like certain organisms, are sometimes what we need at the time to get from point A to point C. All of the so-called missing links are the unstable intermediaries that acted to reach a new level of stable complexity. These organisms (and ideas) are like scaffolding—necessary to finish the job at hand, but nothing we should expect to stay up just because we built it. Alchemy was an important intermediate to get us into physics and chemistry—but it would be foolish to continue taking alchemy seriously as a science just because humans at some point came up with the idea. We are wrong about a great many things—as humans, cultures, and individuals—and it is no defense of human dignity to insist that we should support things that are wrong just because someone somewhere at some time believed in it. Just because dinosaurs once existed, it does not mean that they deserve to exist now. The theropod dinosaurs at the very least seemed to be a highly stable transition state to get us to birds. The sauropods, on the other hand, were just a dead end. The same is true of ideas, including laws. In the United States there are many laws on the books that were once necessary, but are no longer. There are those who argue that affirmative action had its time and place, but that it has certainly outlived its usefulness in the United States of the 21st Century. Of course, when politicians won't even get rid of laws that are clearly destructive because those laws match some wrong-headed ideal, it is a lot to expect politicians to get rid of laws that did once have their desired effect, even if they no longer do. The problem with laws is their permanence. In the United States, this could be solved with a sunset amendment to the Constitution, which would make it so that each law passed by the federal government (other than the Constitution, of course, which should provide a level of legal stability for the U.S.) would necessarily have to be voted on again in, say, ten years. Any laws that were truly good enough to continue to exist would be voted for again and again, while out of date laws would be voted down, or allowed to just fade away into the sunset. There would still be some bad laws that would get voted for again and again, but many more would be eliminated. And perhaps we would get rid of tons of scaffolding. Thus would our laws come to resemble more natural systems, on the edge of order and chaos, and thus become a better set of laws.

200.

While all ideas are emergent from the human brain, and act as strange attractors for human social systems when the ideas are attractive enough, ideas are themselves incapable of being emergent. It seems emergence into the next level of complexity is a special, and rare, occurrence. Since each emergence into new levels of complexity have resulted in more materiality, let me suggest that whatever will emerge from the human will be even more material, and have more material reality, than do humans. Perhaps at some point, what will eventually emerge from us that itself has emergence will thus be technological—though one cannot exclude it also having various cultural elements as well, such as language, artistic-technological creativity, etc.—and will integrate human-type minds, or perhaps more complex minds, into it in a neotenous fashion to leap forward in complexity. Perhaps that technology will directly integrate humans into it, or perhaps there will be a development where direct integration will not be necessary. I will say that whatever emerges beyond the human and into the metahuman and beyond, in potential combinations with technology, there will be far fewer of them than there are humans, in the same way that there are far fewer humans than there are, or ever was, living organisms, just as there are fewer living organisms than there are molecules, and fewer molecules than there are atoms, and fewer atoms than there is overall energy in the universe. Thus it is likely that the new emergent system will include humans, but not all humans (not all vertebrates are human, and not all animals are vertebrates, etc.). Few humans will emerge into becoming metahumans. In the same way that chemistry tinkered with itself until it gave rise to life and, eventually, humans, the human mind will tinker with its own strange attractors until it gives rise to a new emergent system. The new system will at first be at some sort of equilibrium, but then will create a symmetry between that new equilibrium and a new far-from-equilibrium state that will be even more creative of even more new things than humans could ever even dream of. It is hard for human thinkers to imagine what that new system will look like or act like, but those who have emerged into the new level of complexity will be doing things humans understand, and will be doing other things that from a human perspective are unimaginable and can only be understood as miracles.

On Health and the Holy

201.

The words holy, whole, hale, and health are etymologically connected—Old English *halig*, *hal*, and *hÆlth*, respectively, are all related to *hal*. Thus are they conceptually connected. To see the world as holy is to see the world as whole—it is to have the world "appear infinite and holy, whereas it now appears finite and corrupt" (William Blake, "The Marriage of Heaven and Hell"). The word holistic comes from the Greek *holos*, whole. Thus the holy is holistic—God is holy because He encompasses all. Perhaps one could even say that we can recognize the divine only when we come to see the world as a whole, when we see the universe as universal. When we can come

To see a World in a grain of sand,
And a Heaven in a wild flower,
Hold Infinity in the palm of your hand,
And Eternity in an hour. (William Blake, "Auguries of Innocence")

202.

Health and hale are the same. To have health is to have wholeness. To be hale is to be healthy—whole and complete. To make healthy is to make whole again. One is healed through medication. Ideally, one would rather maintain health than have to withstand the ravages of medication (pharmaceutical comes from the Greek *pharmakon*, which means both poison and medicine—as "drug" does today), though medication is necessary to stave off disease. This is the purpose of Plato's *pharmakon*, to stave off disease. For disease is the opposite of health.

203.

The Modern Era, which we are still in, though we may be at the end of it, began with Descartes' solipsistic splitting of man in two—body and soul. It was a necessary division for the development of modern science (which Descartes all but admits to—the di-vision is so the Church will tend to the soul, while the body is left alone, to be tended to by scientists such as Descartes), but it was certainly an unholy division (as all divisions are, by definition). Kant deepened this division. Hegel tried to mend it through philosophical synthesis. Marx tried to mend it by recommending the overthrow of half the world, making it wholly Proletarian. Nietzsche responded to Hegel by dividing the world up even more—for him, humans are not divided into body and soul, but are instead a series of masks. With post-modernism, the division is complete: men and women, multiculturalism, radical Cartesian solipsism divide us up more and more. Any universality is denied. A necessary development—and not without its truth (I am aware of the irony of using the word truth, which comes from the Old English *treowth*, related to the word troth, from which we get the word betrothed, to speak of an idea that is more interested in divorce than betrothal).

But it is precisely as unholy as one can get. More, the denial of universality has also brought about the near-death of the university.

204.

The deep divisions fostered by postmodernism came about because of a view that grand narratives, attempts to universalize, and seeing the world as holistic created the problems of the 20th Century. The Marxist grand narrative gave us the gulag of the Soviet Union, the massacres of the Khmer Rouge in Cambodia, and many other slaughters. Governments who embraced Marxist philosophy seemed to consistently slaughter their own people—something not too surprising once we realize how deeply dehumanizing Marxism truly is. We looked at history after the Holocaust, saw the grand narrative of Christianity had itself promoted the killing of Jews in the past—particularly in the Inquisition—and concluded that it too was dangerous. One could also mention The Terror of the French Revolution. What did Marxism, Medieval Christianity, and the French Revolution have in common? One thing was that they were all grand narratives. Thus, the logic goes, it must be grand narratives which are bad. And what do grand narratives do? They see the world as a whole, which must be encompassed by their ideology. To make the world a whole, it must be placed under their one ideology. Thus, holistic world views were seen as bad—thus were they, and holiness, rejected. The path to Heaven—whether that heaven was celestial or earthly—seemed to lead us straight into Hell. Perhaps in part the rejection of holding a holistic view came about because it is related to the holy, and the holy has been rejected. To the extent that wisdom is the ability to see the unity of the world—meaning wisdom is the ability to see the world as holy—wisdom was also rejected as impossible, perhaps even undesirable.

205.

The error in this way of thinking come from the error made in seeing Communism, Christianity, or the ideals of the French Revolution as interested in seeing the world as holistic. None of them saw the world as holistic, as holy—they instead wanted to make the world whole, under their particular umbrellas. They too fostered divisions—there were enemies who had to be either converted or killed in order that the world could be made holy. "For the cherub with his flaming sword is hereby commanded to leave his guard at tree of life; and when he does, the whole creation will be consumed and appear infinite and holy" (Blake "The Marriage of Heaven and Hell"). Now Blake here uses the word "appear." None of them saw the world as holy. It had to be made so—through conflagration, if necessary. Postmodern thought, by dividing the world even more, does not help us to see the world as holy—quite the contrary. However, by insisting on equality among the various parts—among individuals, among cultures, among religions, among any number of groups of individuals—postmodern thought may ironically make it now possible to see the world as being, rather than needing to become, holy.

206.

It may seem ironic to suggest that only by reaching the most severe of divisions—seeing the world as eminently unholy—that we can finally come to see that the world is in fact holy, but I am not being ironic. To see the world as holy is not to see everything in the world as equal in an egalitarian sense. There are hierarch-ies. To see the world as holy is to understand how everything fits into the world as a whole. It is to see the world as an immense organism, and to care for its health. An organism is made up of systems, organs, tissues, cells, organelles, and various biomolecules. For one group to want to envelop the entire world in one way of thinking, believing, viewing the world, would be the same as one cell wanting to envelop the entire organism in that one type of cell. We have a word for cells that want to do that: cancer. The postmodernists have mistaken cancer for the whole organism. Cancer must be fought, not mistaken for the animal it is in. That is the only way one can have a healthy organism—and it is the only way to have a holy world. Like a healthy organism, a holy world is complex. Like a healthy organism, a holy world has smoothly working parts in communication with each other through clear rules that proscribe what each part needs to do for the whole to work well. Like a healthy organism, this cannot come from any centralized authority—there is no one control cell in the body, and the brain must have the lungs just as much as the lungs must have the brain. A holy world is like a healthy organism.

207.

In Negative Theology, one comes to know what God is by figuring out what God is not. Aristotle says that if you are not sure if something is good, try to figure out what is bad, and you can then deduce that what is good is its opposite.

208.

The following are unhealthy:
1. Overeating, including eating a high percentage of foods with low nutritional value, while remaining inactive (not exercising)
2. Either stagnation or change without continuity—both create instability
3. Stress and anxiety—which comes about from not realizing that there are parts of the world that one cannot control, and can lead to anger at those very things
4. Hatred—aside from raising the blood pressure, it can cause one to act in ways that would be unhealthy for the object of our hatred
5. Cancer—already discussed
6. Excess—including the excess of moderation
7. Shackles—prevents sufficient movement, equating to lack of exercise
8. Pollution—it can lead to any number of diseases
9. An overly-clean environment—it can prevent our immune systems from developing properly, making us more susceptible to diseases, especially autoimmune diseases
10. Suicide—inherently and obviously unhealthy

11. Isolation—loneliness can lead to depression, which depresses the immune system
12. Ignorance—either of the world or of oneself, as one cannot maintain one's health if one is ignorant of what can harm it or improve it

209.

This leads one to posit the following are healthy:
1. Exercise, with a diet proper to the amount of exercise and of high nutritional value
2. Change with continuity
3. Realizing that there are parts of the world that one cannot control, thus reducing stress and anxiety
4. Love
5. Keeping the body in hierarchical harmony
6. Moderation in everything, including moderation—remembering that moderation is an extreme in the same way that life is an extreme state of organic chemistry
7. Freedom—remembering that freedom does not equate to a lack of rules, but is rather what is achieved through playing by the best rules
8. Cleanliness (which, as the saying goes, is next to Godliness, meaning it is holy)
9. A non-sanitized world—a world without dirt is a world that makes unhealthy organisms
10. Love of one's own life and self
11. Friends
12. Knowledge—including self-knowledge

210.

One can make a similar list of what makes for a healthy mind:
1. Taking in healthy information—good art, music, literature, philosophy, the sciences, etc.—with sufficient exercise of the mind through thought, discussion, and writing
2. Change with continuity
3. Realizing that there are parts of the world that one cannot control, thus reducing stress and anxiety, which can negatively affect the mind as well as the body
4. Love
5. Having a variety of inputs—obsession with one thing alone is a kind of mental cancer
6. Moderation in everything, including moderation—moderation of reading, of rigorous thought, sexual thoughts, work, play, physical activity, etc.
7. Freedom of thought—we must not think in shackles, but with flexible rules
8. Cleanliness of thought—this does not necessarily mean what it has traditionally meant in the West and other cultures, though it can certainly contain some elements; thinking about sex, for example, is in and of itself not unclean

9. Realization that we do not and cannot live in a sanitized world, as that stops thought and creativity
10. Love of thinking
11. Friends—as Aristotle says in *Rhetoric*, "a wide circle of friends, a virtuous circle of friends," and, I would add, a mentally stimulating circle of friends
12. Knowledge—including self knowledge—with the goal of wisdom

211.

A holy world is one that parallels the healthy body and the healthy mind, and will have the above qualities, including moderation in everything (i.e., it will be a just world), freedom to act and speak (which does not infringe on others' freedom to act and speak), love, friendship, and beauty. A holy world is a beautiful world, both having variety in unity, unity in variety, complexity, and fluid hierarchy that is self-similar regardless of scale. All of the parts, living in love and friendship (which does not exclude healthy competition, such as we find in sports and in free trade), living in a complex dynamic with each other, living as individuals in various communities, many of which overlap and are nested within other communities, must be self-similar to have a holy world.

212.

In Greek, *kalon* means beauty, but it can also mean honorable or noble—and *kala* can mean either things that are beautiful or things that are morally good. In the *Rhetoric*, Aristotle says, "Now *kalon* describes whatever, through being chosen for itself, is praiseworthy or whatever, through being good [*agathon*], is pleasant because it is good [*agathon*]. If this, then, is the *kalon*, then virtue is necessarily *kalon*; for it is praiseworthy because of being good [*agathon*]" (79). Elaine Scarry points out that in English too there is a connection between beauty and the good (the just), when she points out that to say that something is fair is to say that it is either beautiful or that it is just. In Greek and in English, the beautiful and the good are connected. If a holy world is a beautiful world, it is a good and just world as well. As Heraclitus said, "For god all things are fair and just, but men have taken some things as unjust, others as just" (LXVIII). The key here is that we see the world itself as just—not the actions of each and every individual. The world is itself justified and cannot itself be unjust. This is consistent with the teachings of any religion that sees the world as having been created by God or the gods—how could a fair and just god create a world that was itself unjust? For example, Genesis says of God's creation that it was good. And if theistic religions are rejected, how can the world itself be unjust? To say it is unjust is to give it anthropomorphic qualities. It is people who have taken some things as being just, others as unjust—but the world itself is self-justified. Those who do not see the world as just are those who do not see the world as holy—often they are the same people who think the only way the world can be justified is if the world is made holy through the transformation of it into a perfect mirror of themselves. But we have seen that a world made up of only one worldview is a cancerous world—and the world, as a cancerous organism, will die. An organism cannot consist of one type of cell—that is the unhealthiest of organisms. And a world having only one world view is the unhealthiest of worlds. In the same way that a healthy body consists of a variety of cells that are variations of the same theme coded by identical DNA, a healthy world consists of a variety of peoples that are variations of the same themes coded for by our being human and sharing the same cultural universals.

213.

Beauty is also related to health—as we can see in the beauty we find in nature. Healthy plants produce the most beautiful flowers. Healthy peacocks produce the largest, most symmetrical, most colorful feathers. Healthy gobies and other territorial reef fish have the brightest colors. All of this natural beauty is the advertisement of health to the opposite sex and to rivals of the same sex. The healthiest human bodies (neither overweight nor super model thin) are the most beautiful. Thus is there also a relationship between health, beauty, and sex. If beauty can thus be equated to health, we can see that beauty is again equated to the holy. And we can see too that sex in-and-of-itself is not and cannot be unholy, as it is connected (but not equivalent) to beauty. Sex allows for the faster creation of variety, thus contributing to the beauty within a species.

214.

A holy world is a whole world. It is a healthy world. It is a good and just world. It is a complex world. It is a world of individuals in community. It is a beautiful world. But is it a possible world? I have already given the answer: the world is itself already holy. We just have to learn to see it as holy. That is how we will heal the world. And, as we do, we will become less and less likely to want or try to eliminate those who disagree with us—until we are all in agreement on this one issue, as all the cells in an organism are in agreement on the one issue that they must work together for the health of the whole, even as each performs its own function. Thus, the world will become more and more holy in our eyes. In works of tragedy, *nomos* (convention, human law, naming; from which we get the words nomad and nomenclature, and which is the changing and changeable aspect of the world) comes into conflict with *physis* (or nature; from which we get the word physics). That is the position we are now in. When we get *nomos* to map onto *physis* (Heraclitus calls this conjunction *logos*—which can be translated as saying, speech, discourse, word, account, explanation, reason, principle, collection, enumeration, ratio, proportion), we will see the world as it truly is: holy. Just because the world is not perfect according to human (mis)calculation and (mis)understanding, that does not at all negate its holiness. To the extent that the world is not holy, it is because we make it so—too often in our attempts to simplify a world we mistakenly believe to be complex-disorderly rather than to make it more complex-orderly.

215.

On the Holy

Where lies the holy in the modern world?
It lies in Blake's world in a grain of sand –
It lies, and lies like the truth, in patterns
Like self-organized rings of rocks barren
Arctic fields create. It lies in the branch
Of every tree and species, leafing out
From the known into the unknown. It lies
In every song, painting and rhythmic verse.
We have looked at every leaf and petal,
At the bark and at the wood, every cell
And strand of DNA is now known –
And we have forgotten that all of this
Was once a tree that gave us shade and filled
The air with delicate sweetness and held
The grains of sand against its roots to hold
The ground in place, even as that ground moves
And changes in tiny ways we refuse
To see. In this we can see the holy.
This is where it lies, now and forever,
On the edge of order and wild chaos,
Where the infinite holds in the finite,
Where we, ourselves holy, have always lived.

Paradox

216.
Complexity in the world comes about through the universe attempting to resolve paradox—only to create even more paradoxes. The simultaneous retention while resolving of paradox (which is itself a paradox) is how we get emergence into new levels of complexity—which contain their own paradoxes.

217.
Information is a paradox—it is inform, and yet gives form. When information is pure, it ceases to give information.

218.
A paradox is not a contradiction. With a contradiction, it is impossible for both A and not A to be true. But with a paradox, it is not impossible for both to be true—both are true.

219.
A paradox is not mere opposites. Opposed pairs "are in a way the same and different" (Aristotle, *Physics Book I*, 189a, p. 12). Hot and cold are opposites—but one cannot have something that is paradoxically hot and cold. There is a continuum from hot to cold, and an object can become hotter or colder by moving along that continuum.

220.
However, a system can become both hotter and colder, by becoming locally hotter and colder—or textured (Koen DePryck).

221.
Something being both straight and curved simultaneously is a paradox. Something being both finite and infinite simultaneously is a paradox. Something being orderly and disorderly simultaneously is a paradox. Paradox is the agonal unification of opposites.

222.
Quantum particle-waves are both particles and waves simultaneously. These are different in a way that is not really the same, in the way we typically think of things being the same—except for the fact that we are talking about the same particle-wave. Since something cannot be both an object and a wave simultaneously—and yet things such as electrons are indeed particles and waves simultaneously—we have here a paradox.

223.

A thing is beautiful if it is paradoxical. Beauty has or contains the following features:
- Complexity within Simplicity
- Digital-Analog
- Emergent from Conflict
- Evolutionary (changes over time)
- Generative and Creative
- Hierarchical Organization
- Play (a nonserious thing done seriously)
- Reflexivity or Feedback
- Rhythmicity
- Rule-Based
- Scalar Self-Similarity
- Time-Bound
- Unity in Multiplicity

These are also features of the universe as a whole. Christian Fuchs lists the following features as aspects of self-organization:
- Emergence
- Complexity
- Cohesion (digital-analog)
- Openness
- Bottom-up-Emergence
- Downward Causation
- Non-linearity
- Feedback Loops, Circular Causality
- Information
- Relative Chance
- Hierarchy
- Globalization and Localization
- Unity in Plurality (Generality and Specificity)

And for Emergence, he lists the following aspects:
- Synergism (productive interaction between parts)
- Novelty
- Irreducibility
- Unpredictability
- Coherence/Correlation
- Historicity

If we compare the lists, we can see there is a correlation between self-organizing complex systems and beauty. Each have the same attributes. "Cognition, co-operation and communication are phenomena that can be found in different forms in all self-organizing systems. All self-organizing systems are information-generating systems. Information is a relationship that exists as a relationship between specific organisational units of matter" (Fuchs). All beautiful objects are information-generating systems. And to the extent that something is a self-organizing system, it is beautiful.

224.

Quantum physics predicts the creation of neutral atoms through the combination of electrons, protons, and neutrons. But it also predicts, in typical conflict with this rule (the exceptions being the noble gases, like helium), the increased stability of a full electron shell—with decreasing stability the fewer the electrons in the shell. This is a paradox. There are no ions in quantum physics—they arise from chemical interactions between atoms, whose interactions make use of unstable electron orbits, stabilizing them with the chemical interaction, transferring the single unstable electron from the donor (often metal) atom to the electron shell of the acceptor (nonmetal) atom, whose outer electron shell is stabilized by having the optimal number of electrons. A sodium atom is stabilized by donating its outer electron to the chlorine atom, stabilizing it. If one merely knew of quantum physics, and not chemistry, could one predict such an interaction? The paradox remains at the atomic level, and, by remaining, is resolved by atoms undergoing chemical reactions.

225.

Quantum elements are inherently both digital and analog: "Quanta are by definition discrete, and their states can be mapped directly onto the states of qubits without approximation. But qubits are also continuous, because of their wave nature; their states can be continuous superpositions" (Lloyd, 152). For this reason, quantum mechanics allows for the ability to register both 1 and 0, two different commands, simultaneously, in quantum parallelism. "The ability to do two things at once arises form the wave nature of quantum mechanics. Each possible state of a quantum system corresponds to a wave, and waves can be superimposed." And "superposing waves results in qualitatively new and richer phenomena" (137).

226.

Heisenberg's uncertainty principle says that we cannot know with simultaneous high accuracy the position and velocity of a quantum mechanical object. Less well known is the fact that "Another one of the pairs of complementary attributes in quantum mechanics is energy and time" (Seife, *Decoding the Universe,* 204). If increasing complexity results in more energy, this could explain how we get increasing complexity of time experience (including the experience of more time at each level) as well, as Fraser claims.

227.

Without cooperation, nothing could get together to create complexity. Without conflict, no tensions would exist to lead to creativity. The universe has evolved greater complexity and greater creativity—meaning cooperation and conflict (love and strife; *eros* and *eris*) are necessarily present. Everything in the world is both in conflict and in cooperation. Both, simultaneously.

228.

All bonds are created through the combination of attraction and repulsion—chemical bonds, atomic bonds, love bonds, and economic bonds. Love and strife make the world.

229.

Chaos theory—both order and disorder, simultaneously. Mathematical fractals—finite space surrounded by an infinite border. The world is chaotic. The world is fractal (statistically fractal, though, and limited by the Planck length).

230.

The golden mean ratio, the very principle of growth, is a paradox. As a mean, it is symmetrical. Yet, it grows at a constant rate, in the ratio of 1:1.618 at each iteration. Thus, it is asymmetrical. Being symmetrical and asymmetrical both, it is a paradox. Further, this ratio contains in it an irrational number: 1.618 . . . Thus, the golden mean ratio is both rational and irrational. Thus, it is doubly paradoxical. No wonder it creates so much, and so much beauty.

231.

There are an infinite number of numbers. If we placed all the numbers on a number line, we would have an infinitely long number line (remembering that this is math, and thus an idealization). As Galileo pointed out, if we remove every odd number, we will have both half the number of whole numbers, and an infinite number of them. One could also have every third number, every tenth number, or the squares of all whole numbers, each on their own infinitely long number lines, and containing an infinite number of those numbers that are nonetheless less than the infinite number of whole numbers. Bertrand Russell pointed out that this is problematic at best, since what we have are different sets of sets, and there are an infinite number of those kinds of infinite sets. The problem arises because then you have a set of those infinite sets, which would then contain itself. It becomes self-referential, which is itself a paradox. An example would be: "This sentence is false." Galileo realized it was a paradox. Cantor showed otherwise, that both were aleph 0, but discovered another two different kinds of infinities—the one including all whole numbers being twice as infinite as the one including only even numbers.

232.

Aleph 1, which is greater than aleph 0, is another kind of infinity. Consider the space between zero and one on the number line. In that space are an infinite number of fractions. One could start with 0.1, 0.11, 0.111, 0.1111, etc.—and then move on to 0.2, 0.3, etc., as well as other repeating decimals, such as 0.121212 . . . , 0.112112112 . . . , etc. These are the rational numbers, of which there are infinitely many. There are also in-finitely many irrational numbers, or nonrepeating decimals (Pi = 3.1415 . . . , the golden mean ratio = 1.618 . . . , the inverse natural logarithm, e = 2.71828 . . . , etc.). Which means that an infinite number of numbers fit in the space on the number line between zero and one (and between each pair of whole numbers). So infinity both fits within any given finite space between two numbers on a number line, and stretches out infinitely far in either direction on the number line. This is the paradox of infinity.

233.

Infinity is impossible in a finite universe (the universe is finitely large, and the Planck length is as finitely small as one can get in the universe). And yet we humans can conceive of not only infinity, but of various paradoxical kinds of infinities. This, too, is a paradox. It is also why we have to be careful not to mistake our concepts for reality. Mathematical fractals are beautiful, but they are only precise approximations of real fractals. Mathematical fractals have infinite depth of self-similarity. Real fractals have finite depth of self-similarity. That does not prevent both from being kinds of fractals.

234.

But I have also said that the universe does in fact have an infinite border. Time both creates and solves this paradox. It is a paradox of the real and the ideal.

235.

In quantum string theory—one of the candidates for unifying gravitation with the strong nuclear, weak nuclear, and electromagnetic forces—everything in the universe is ultimately made up of tiny vibrating strings that "play the familiar medley of particles as if they were musical notes" (Amanda Gefter, *Scientific American*, Dec. 2002, 40). These strings may be either linear or circular—or both. How both? There are two possibilities. Circular strings could give rise to one form of particle-wave, while linear strings could give rise to electrons, neutrons, and protons, which can interact in more complex ways. The other possibility (and it may contain the first one in it, and vice versa) is that there is a recursive element to strings—that they are simultaneously linear and circular. Recursive linear strings vibrate and interact (provide information to each other) to give rise to more complex atomic systems, which includes chemical systems. I suspect the latter is the case—that paradox goes all the way down to strings (which would make them self-similar to everything they have given rise to). Strings too are self-referential.

236.

Another form of recursive linearity giving rise to complexity is in the genetic material of organisms. Prokaryotic cells have circular strands of DNA, while more complex eukaryotic cells have linear strands of DNA. DNA itself looks linear (but wavy) from the side, but circular from the top, due to its helical structure. This suggests, if there is scalar similarity, my first theory of strings. But both forms of DNA are recursive in their cellular interactions, suggesting the second. This scalar similarity at the organismal level suggests how understanding organisms can help us understand strings. Relative to a "3-D"cell, DNA is a "1-D" string. This string interacts with itself through other kinds of strings (RNA, proteins) to give rise to a higher-dimension reality—life. These "strings" are in one sense 3-D; but in another sense, they have far more dimensions—in the number of genes, regulatory sites, etc., in the DNA. DNA can have hundreds to tens of thousands of dimensions—which interact to give rise to cells more complex than is the DNA itself. In the same way, strings are 1-D relative to the "4-D" universe, but also 10.54-D systems giving rise to the poly-dimensional universe. Paradoxical recursive linearity gives rise to higher-order complex systems.

237.

It seems there is an element to the universe that makes recursive linear systems give rise to more complex systems. This element may be the string-foundation of the universe, which is scalarly projected into higher levels of complexity. Information-containing strings interact with themselves in complex ways to give rise to complex systems. Linear strings gave rise to complex quantum physics, including chemistry; certain forms of chemistry gave rise to linear strings of genetic material which gave rise to complex organisms; certain organisms with complex brains gave rise to linear strings of words, which gave rise to the full flowering of human culture, including art and literature and technology. This is a universe which is self-similar to at least three scales of recursive linear elements giving rise to complex nonlinear systems.

238.

The mathematics of emergence says that if you have two identical systems, A and A' (both A and A' being sets), then A=A' and A>A' when A becomes a self-assembling system with emergent properties. The difference? Density. When spacetime becomes more dense, A becomes greater than A', even though A is still equal to A'. This is a paradox—and it is the very foundational paradox of the universe.

239.

Without limits and constraints, there is no freedom. With limits and constraints, there is no freedom. Rules are limits and constraints, and so are laws. Without them, there is no self- organization, there is no emergence, there is no creativity or innovation. With them, there is only entropy, simplicity, destruction, and death. Without them, there is only pure energy—if there is even that. With them, there is only pure energy—if there is even that.

240.

To understand anything, it has to be divided, clarified, and simplified. Otherwise the mind will be filled with nothing but muddlement. This is, after all, how the world is fundamentally constructed. We know the world through reductionism. This is knowledge.

Paradox

241.

To understand anything, it has to be unified, mixed together, complexified. Otherwise the mind will be filled with nothing but nonsense. This is, after all, how the world is fundamentally constructed. We know the world is complex. This is wisdom.

242.

Probability is subjective and objective simultaneously.

243.

From whence do paradoxes arise? Self-referentiality. "Gödel showed that the capacity for self-reference leads automatically to paradoxes in logic; the British mathematician Alan Turing showed that self-reference leads to uncomputability in computers" (Lloyd, 35). J. T. Fraser suggests that emergence into new levels of complexity result in the resolution of these paradoxes (though it is a resolution which keeps the paradox in existence at the lower level). All systems governed by rules are logical systems, and a system governed by rules can never be complete, as Gödel showed in his Incompleteness Theorem, since

> Above a certain level of complexity, there are intrinsic limits to a logical system, if that system is consistent. There will always be true statements which can neither be shown to be true nor proved to be false within the confines of the system, using the axioms or rules of the system. Moving outside the original system, enlarging it by adding new axioms or rules may make the statement provable, but within this wider metasystem there would be other statements that could not be proved. (Campbell, 109)

Each new level of complexity is precisely this kind of metasystem. Paradoxes are solved, but more paradoxes are created.

> It is customary to assign our own unpredictable behavior and that of other humans to irrationality: were we to behave rationally, we reason, the world would be more predictable. In fact, it is just when we behave rationally, moving logically, like a computer, from step to step, that our behavior becomes provably *un*predictable. Rationality combines with self-reference to make our actions intrinsically paradoxical and uncertain. (Lloyd, 36)

On the Creation of Complexity in the Universe

244.

The current laws of physics do not explain the complexity of the universe. Thus, we must be missing something. This is the argument Stuart Kauffman makes in his *Investigations*, and he lays out several suggestions that create an incomplete picture perfect for further investigations. I would like to propose a model that could explain the creation of complexity in the universe—not just more things, but the emergence of more complex systems of things, that nonetheless have fractal self-similarity to the less complex systems below them, and which make them. These are two different kinds of things that the universe manages to create. The creation of more, and more complex, things in the universe is proposed by Kauffman too in *Investigations* to perhaps be a better way of gauging the passage of time than is the current way of doing so through using the increase in entropy in the universe. But we will have to undertake our own investigations here to see if that model makes sense, or if there is a correlation between it and entropy increase as a measure of the arrow of time.

245.

Prior to the big bang, there was nothing—or, perfect symmetry. This symmetry broke, to create the singularity that gave rise to the big bang, which gave rise to all the energy in the universe. The action that gave rise to the universe was thus symmetry-breaking (branching), suggesting symmetry-breaking is the first, most fundamental law of physics, which gave rise to everything in the universe.

246.

But how? And how did that universe settle on the particular laws of physics which would give rise to atoms, molecules, life, and intelligence that could ask these kinds of questions? And what, after all, is symmetry-breaking? Robert Laughlin, in *A Different Universe*, spells out what symmetry-breaking is quite clearly:

> The idea of symmetry breaking is simple: matter collectively and spontaneously acquires a property or preference not present in the underlying rules themselves. For example, when atoms order into a crystal, they acquire preferred positions, even though there was nothing preferred about these positions before the crystal formed. When a piece of iron becomes magnetic, the magnetism spontaneously selects a direction in which to point. These effects are important because they prove that organizational principles can give primitive matter a mind of its own and empower it to make decisions. WE say that the matter makes the decision "at random"—meaning on the basis of some otherwise insignificant initial condition or external influence—but that does not quite capture the essence of the matter. Once the decision is made, it becomes "real" and there is nothing random about it anymore. Symmetry breaking provides a simple, convincing example of how

nature can become richly complex all on its own despite having underlying rules that are simple. (44)

It is important to understand this simple idea if we are to understand how more and more complex things evolved in the universe, to make the universe full of the kinds of things we observe in it (including ourselves).

247.

We know that the universe began in a burst of pure energy, then set out on a series of bifurcations. Gravity separated out first from the GUT force, then the strong nuclear force separated out from the electroweak force, then the electroweak force bifurcated into the electromagnetic force and the weak nuclear force. An atom thus exists on the borderlands of the strong nuclear, weak nuclear, and electromagnetic forces, which work together to hold the various elements of an atom together. This makes each atom a small complex system, stabilized by each of the attractors in the atom-system. Other laws of the universe act as less complex attractors: the second law of thermodynamics, for example, is a simple attractor pulling closed thermodynamic systems toward maximum entropy. Gravity is another simple attractor that pulls mass toward the center of that mass. Gravity around a single star is a simple attractor—it pulls objects toward the star. However, centrifugal force also plays a part as an attractor, making the planets orbit the sun in elliptical orbits rather than being pulled into the sun itself. Things become more complex when we deal with more equally massive stars. Two stars orbit each other in a predictable fashion, but three stars, with their own strong basins of attraction, orbit each other in a more chaotic fashion, which cannot be mathematically predicted. A more interesting problem, which helps illustrate the overall problem, would be if there were a planet in orbit in conjunction with two stars of equal size. The two stars would create equal basins of attraction, and the planet, if it were positioned halfway between the two stars, would be on the border between the two—the bifurcation point. So long as the planet remains on the border between the two stars, its orbital behavior will be highly chaotic and unpredictable. It would become predictable only if it managed to fall into either one or the other basins—in which case, the planet would fall into a more predictable orbit around one star.

248.

Here we are dealing with very simple attractors. But what happens when we are dealing with several attractors of different strengths and sizes? This seems to be the situation with atoms. Each atom has the three atomic forces holding it together. Further, atoms are made of elements with mass and electrical charges. Each of these are attractors. In the case of an atom more complex than hydrogen, we have nuclei with several positive charges which repel each other, yet are held together by the strong and weak nuclear forces, and is further stabilized by a negatively charged electron shell surrounding the nucleus, which have their own repulsion in all being negatively charged. This creates a complex interaction that is the atom. One could perhaps see each atom as a strange attractor, and each different kind of atom as a different kind of strange attractor, since an atomic

On the Creation of Complexity in the Universe 93

system is made up of forces that amplify differences and forces that dampen differences, meaning each atom is in fact a nonlinear system.

249.

Stuart Kauffman shows that for any system with a certain number of components (N), that system will have $2^{n/2}$ possible states within the system, but only N/e number of cycles, or possible basins of attraction, where *e* is the inverse natural logarithm (e=2.7182818284 . . .)

> Thus a system containing 200 elements would have only about 74 alternative asymptotic patterns of behavior. More strikingly, a system containing 10,000 elements and chaotic attractors with median lengths on the order of 2^{5000} would harbor only about 3700 alternative attractors. This is already an interesting intimation of order even in extremely complex disordered systems. (*Origins of Order*, 194)

Kauffman then shows that such systems are even more organized, since for a complex system:

> The expected median state cycle length is about $\sqrt{N}$. That is, the number of states on an attractor scales as the square root of the number of elements. A Boolean network with 10,000 elements which was utterly random within the constraint that each element is regulated by only two elements would therefore have a state space of $2^{10,000} = 10^{3000}$ but would settle down and cycle recurrently among a mere $\sqrt{10,000} = 100$ states. . . . A system of 10,000 elements which localizes its dynamical behavior to 100 states has restricted itself to 10^{-2998} parts of its entire state space. Here is spontaneous order indeed. . . . The number of state cycle attractors is also about $\sqrt{N}$. Therefore, a random Boolean network with 10,000 elements would be expected to have on the order of 100 alternative attractors. A system with 100,000 elements, comparable to the human genome, would have about 317 alternative asymptotic attractors (201).

Kauffman suggests we should understand each kind of cell as a different strange attractor. As it turns out, 317 is about how many kinds of cells one finds in the human body. More importantly, systems with very large numbers of elements can and do have a very small number of ways of organizing themselves, though the number of ways of expressing those rules may be astronomical. For a system with N=200, the median cycle length, or possible states per system, is $2^{100} \cong 10^{30}$, and "At a microsecond per state transition, it would require about a billion times the age of the universe to traverse the attractor" (Kauffman, 194). And that is for a tiny system with only 200 elements. Yet the actual ways such a system would be expressed would be only 47. There would be 47 general forms, with 10^{30} specific forms. These strange attractors (though not the specific numbers I have used as examples, of course) are the different species of animals the "zoological system" can create; the median cycle length is the number of particular individuals that could be generated.

250.

Seth Lloyd says that for a quantum computer, "the number of quantum searches required to locate what you're looking for is the square root of the number of places in which it could be" (143). Lloyd also says that the universe itself is a quantum computer: "since the universe registers and processes information like a quantum computer and is observationally indistinguishable from a quantum computer, then it *is* a quantum computer" (154). The question is, is this square root the same as Kauffman's square root from above? If so, we have a quantum mechanical description of the origins of strange attractors—and strange attractors are necessary if we are to have complex systems.

251.

The universe started off with mostly hydrogen, some helium, and a trace amount of lithium—so where did the rest come from? Attractors increase in number when a given system is stressed. When energy is added to a system, that energy can amplify the differences, resulting in more types of behavior. Thus, we get bifurcations. Thus we get bios, or creativity. Gravity acted as an attractor that pulled atoms together, pooling them in increasingly larger basins. This caused the atoms to press against each other, putting stress on the atoms, until the energy stripped the atoms of their electrons, and nuclear fusion occurred. Even in the fusion furnaces of stars, however, there is an upper limit to the kinds and numbers of strange attractor systems (nuclei) that can be created. However, stars that explode into supernovae are able to provide that last little burst of energy needed to create even the heaviest of stable atomic nuclei. We are beginning to see why and how we can and did get more complexity in the universe, in seeming (and we should understand by now why it is only "seeming") violation of the second law of thermodynamics. It is only a violation if we see the second law as the most important, the driving law of the universe. However, if we see that the second law is merely one attractor among many in the open universe, we can see quite clearly that there is no violation of the second law. It is only one of many attractors that must be taken into consideration.

252.

In a sense, we have gotten slightly ahead of ourselves. We can get atoms once we have the right laws of physics, so that stable and more complex atoms can evolve. Not just any laws of physics will do in the evolution of such a universe, just as in biology not just any biological feature will do. Rabbits that glow in the dark would not last long in nature (though they might be selected for by humans, as the rabbit genetically engineered to glow in the dark shows). In the same way, quantum elements that could not interact well, or were highly unstable when they did interact, would not survive. Such systems would collapse. In the transition from energy to elementary particles, there was a search through fitness landscapes, each element as it evolved deforming and transforming the fitness landscapes of other elements, until those elements with the smoothest landscapes—meaning they could all interact and recombine well with each other—evolved. And here is where things get interesting. A smooth fitness landscape is a symmetrical fitness landscape—so the universe evolved all of its particles, photons, forces, and constants toward the creation of a new symmetry—one balanced on the edge of particle and wave, and symmetrical between them to the extent that one cannot predict before observing whether the quantum element is a particle or a wave. However, there is an interesting quality of this kind of

On the Creation of Complexity in the Universe

symmetry, and that is that it is also far-from-equilibrium. To be at equilibrium, one must be either entirely particle or entirely wave—entirely solid or entirely fluid. Instead, quantum elements exist on the borderland between the two, making them far-from-equilibrium. If equilibrium is the same as maximum entropy, we see two opposing forces in the universe—a tendency toward equilibrium, and a tendency toward far-from-equilibrium, or edge-of-chaos, states. And edge-of-chaos states keep their fitness landscapes melted and, thus, smooth. Strangely, if these two forces—equilibrium and far-from-equilibrium—are themselves symmetrical, then the universe as a whole would be doubly far-from-equilibrium, as the borderland between the equilibrium basin and the far-from-equilibrium basin would be an edge-of-chaos border where bifurcation, thus creation, would occur. Actually, it would be more than doubly far-from-equilibrium, since the borderlands between these two basins of attraction would be fractal, thus have infinite length, and infinite depth (though encompassing finite spaces, as fractals do), and far-from-equilibrium states would be infinitely generated on the borderlands.

253.

I think it is a mistake to equate equilibrium and entropy. Certainly, in a closed system it makes sense to equate the two. However, the universe as a whole is not a closed system—it is an expanding system. Perhaps what we need to do is look at several definitions of entropy and see if we can make them make sense in light of this difficulty in explaining the existence of more, and more complex, things in a universe with the second law of thermodynamics.

254.

Stuart Kauffman says entropy is defined statistically. Entropy is the state of a system where the amount of energy we would need to use to learn the state of the system equals or is more than the amount of energy we could get out of it in work. In *An Introduction to Information Theory*, John Pierce says that

> In communication theory, entropy is interpreted as average uncertainty or choice, e.g., the average uncertainty as to what symbol the source will produce next or the average choice the source has as to what symbol it will produce next. The entropy of statistical mechanics measures the uncertainty as to which of many states a physical system is actually in. (289)

If this is how we define entropy, then the universe must create more things, because more things mean more choices and, thus, more uncertainty—that is, more uncertainty of what will be chosen to be made, to survive, etc. So entropy is "a measure of choice, the amount of choice the source exercises in selecting the one particular message that is actually transmitted" (105). It measures "the uncertainty of the recipient as to which message will be received, an uncertainty which is resolved on receipt of the message" (105). When we make measurements of the state of a system, we "ask" the system to "tell" us what state it is in. If we continue to have as much (or more) uncertainty at the end of the inquiry as we had before we made the enquiry, the system is at maximum entropy. Here we have a definition of entropy as the measure of uncertainty. The more uncertainty we have about a system, the more entropy the system has.

255.

In *The Making of a New Science*, Stephen Wolfram points out that with complex systems, the calculations we would have to make to predict the state of the system at any point in the future would take as long or longer than if we just waited to see what the system made. This fits both definitions of entropy given above. For complex systems, we would have to put in the same or more amount of energy into calculating the state of the system (ostensibly to get work out of the system), as we would get out of the system by just waiting for it to get to where it is going. I will go a step further: the more complex a system is, the more difficult the calculations, so that our ability to know the state of a system decreases as systems become more complex. In other words, we cannot know what will be created in the adjacent possible, because by the time we could have possibly calculated it, it would have already been created, and a new adjacent possible created. We cannot know what the future state of a system is, nor can we know even the present state of a system—all we can know are the past states of a system, the past products of the system. And as we get more, and more complex, things, the larger and more complex the adjacent possible becomes, and therefore the less predictable the universe becomes. Certainly quantum particles are more predictable than are biological organisms, let alone humans. The uncertainty of a quantum particle (is it a wave or a particle? do we want to know its position or velocity? is it spin up or spin down?) is far less than the uncertainty of what a human will produce next. Could you have predicted the last sentence? Can you predict my next one?

> A message which is one out of ten possible messages conveys a smaller amount of information than a message which is one out of a million possible messages. The entropy of communication theory is a measure of this uncertainty and the uncertainty, or entropy, is taken as the measure of the amount of information conveyed by a message from a source. The more we know about what message the source will produce, the less uncertainty, the less entropy, and the less information.
> (Pierce, 23)

Although a system is unpredictable, we can still learn about its past states. The less certainty we had before, the more information we can get from the system (though it still has to be related to something known—otherwise we still do not have information conveyed, as we cannot relate the new information to what it already known). So "Entropy increases as the number of messages among which the source may choose increases. It also increases as the freedom of choice (or the uncertainty to the recipient) increases and decreases as the freedom of choice and the uncertainty are restricted" (Pierce, 81). As the universe increases in number of objects, it increases in number of choices. This implies that the ever-increasing number of things at ever-increasing complexity is increasing entropy in an open, expanding universe. We have failed to completely see this because we have mistakenly equated entropy and equilibrium. These are systems such that we start and end with unpredictability. However, with complex dynamic systems, we start with unpredictability and end with information (after the fact, of course), though we will remain in the dark as to what state the system will be in in the future. Thus, it seems that it is in being far-from-equilibrium that entropy (as uncertainty) increases in the universe, since it is when a system is in

On the Creation of Complexity in the Universe 97

this state that we have the least certainty, in that we don't know what the system is going to produce next. So, when Kauffman suggests we replace as a measure of time the increasing proliferation of things for increasing entropy, he may be suggesting that we retain the same measure of time.

256.

Any system far-from-equilibrium is creative and self-assembling (Prigogine). Thus is the universe and everything in the universe self-assembling. Quantum elements have interacted in ways (nuclear fusion and chemistry) that created more complex interactions as the elements decohered in their interactions with each other. As quantum elements decohered, the particle-wave symmetry was broken, and there was a proliferation of atomic and molecular forms.

257.

One nuclear fusion/molecular system was our own solar system. On one planet, Earth, there arose a new form of complexity. As it turns out, there is one kind of molecule that is extremely stable: organic molecules. As such, it is in a sense a molecular "generalist" in that it can give rise to an ever-increasing number of forms. There may be an upper limit of (non-organic) chemical reactions of, say, sodium, but there is no limit to the number of chemical reactions of carbon—just add another carbon group, and you get an entirely new molecule. Add another NaCl to a salt crystal, and you get a slightly larger salt crystal that acts the same way as did the slightly smaller one. Add one more CH group to an alcohol, and suddenly what was water-solvent is water-insolvent, or at least noticeably less so. Thus, carbon chemistry is on a smooth fitness landscape, and is thus symmetrical. But this symmetry is, again, far-from-equilibrium. Once carbon chemistry reached an upper limit of complexity, it became far-from-equilibrium, and some of the organic molecules self-organized into life, breaking organic chemistry's symmetry. It did so too by entering into a state between a liquid and a crystal—cells seem to be more like liquid crystals. The cell membrane is an excellent example of this, with proteins imbedded in the phospholipid bilayer, moving (as do each of the phospholipids) within the membrane matrix. There are parts of the cell more liquid, other parts more crystalline, but the cell as a whole can be understood as a liquid crystal. We can see another form of a dynamic system inbetweenness with a hurricane, is a system on the borderland between atmosphere and ocean, something I came to understand when I witnessed my first hurricane in Mississippi and saw the sky bring the ocean ashore. This results in the creation of the strange attractor/eye creating a series of cloud Fibonacci spirals. Those who study hurricanes can tell you that the future states of any hurricane system is all but impossible to predict—including the simplest of information, such as speed and location. A hurricane too is both symmetrical and far-from-equilibrium.

258.

There seems to be an unusual mechanism for leaps into greater complexity (a hurricane cannot become anything more complex because, being on the

98   On the Creation of Complexity in the Universe

borderland between water and air, it is too simple and disordered). Stuart Kauffman makes a suggestion about the relation of life to coherent quantum systems and decoherent chemistry which I at first rejected, but now, for reasons I will soon make clear, consider to be very likely. Kauffman suggests that

> The persistent intermingling of quantum and classical phenomena in a living cell might require quantum coherence, but that coherence is widely doubted at the normal temperatures of cells and organisms. On the other hand, persistent intermingling of quantum and classical phenomena might well occur and not require quantum coherence if the timescale of decoherence is close to or overlaps the timescales of cellular-molecular phenomena. Recent calculations suggest that the timescale of decoherence of a protein in water at room temperature might be on the order of 10^{-19} seconds. Thus, it is interesting that proteins and other organic molecules have modes of motion on timescales over many orders of magnitude, spanning from tens of seconds down to 10^{-16} seconds or less. Thus, the timescale of decoherence is almost the same as the rapid molecular motions in cells. It does not seem totally implausible that cells persistently abide in both the quantum and classical realms, in which the persistently propagating superposition of amplitudes for alternative molecular motions decohere on very rapid timescales and thereby help choose the now classical microstates of proteins and their motions as those proteins couple their coordinated dance with one another to carry out the alternative behaviors that guide a cell in its next set of actions, its adjacent possible. In short, cells may feel their way into the adjacent possible by quantum superpositions of many simultaneous quantum possibilities, which decohere to generate specific classical choices. (*Investigations*, 150)

I consider this idea likely because it is a similar mechanism to how every new level of complexity has been realized in the universe. I propose a sort of universal "neoteny" that gives rise to complexity by creating the symmetries that create far-from-equilibrium conditions that break these symmetries in the creation of a new emergent level. "Neoteny" means "holding on to youth." The "youth" of the quantum world was the particle-less wave-energy that preceded the evolution of the quantum world. Quantum elements "hold on" to their "youth" by existing as particle-waves, while "mature" macrophysical objects are solid particles. Quantum elements have a betweenness. This is what Kauffman is proposing regarding life as existing between coherence and decoherence—coherence as the "youth" of decoherent chemistry.

259.

The first life was prokaryotic. It likely started off as archaebacteria, which live in very extreme environments, and then evolved into eubacteria, which are the bacteria most people are familiar with. Neither create multicellular organisms, so both can only become so complex. Eukaryotic cells are much more complex cells than are prokaryotic cells, and are capable of evolving into complex multicellular organisms. As it turns out, eukaryotes are more closely related to archaebacteria than to eubacteria—though the mitochondria and chloroplasts in eukaryotes evolved from eubacteria that developed a symbiotic relationship with the eukaryotic cells. In this way, a sort of betweenness was created between archaebacteria and eubacteria that was more stable, more symmetrical (able to adapt to more environments and evolve into greater numbers of forms)—and thus, were far from equilibrium. New more rugged fitness landscapes developed for eukaryotes to search and transform into smoother landscapes.

260.

Once you get eukaryotes, you still do not have a direct capability to evolve into multicellular organisms, even if there is now an ability to create polycellular groups. In the line of cells that gave rise to animals, at least, there arose the homeobox (HOX) genes, which are extremely evolutionarily stable. HOX genes from worms, fruit flies, humans, etc. are practically identical. Yet they are necessary for laying out such radically different patterns of development as antennae for fruit flies, and fingers for humans. Insects are strictly segmented, mammals less strictly so, yet practically the same proteins are used to lay out these segments. It seems that, in the same way that carbon provides the kind of stability necessary for the proliferation of molecular forms, HOX genes, by being highly conserved in evolution, provide the kind of stability that is necessary for the proliferation of variation in animals. The HOX genes may thus have been the kinds of genes necessary to evolve from mere poly-celled eukaryotes to true multicellular organisms. The HOX genes, by being so stable, would also allow other genes to be less stable, allowing for more evolutionary experimentation. Further, the stability of the HOX genes would allow them to act as strange attractors for laying out the patterns of animal development and structure. Even a small change in a HOX gene can cause large changes in structure—as evidenced in the familiar examples of fruit flies developing legs where their antennae should be. Most HOX mutations will be fatal—this keeps them stable—but a few may result in positive changes that are nonetheless spectacular in consequence. For example, vertebrates and insects have completely opposite developmental strategies. When the embryonic ball of cells folds in, what becomes the mouth in vertebrates becomes the anus in insects, and vice versa. There was likely a mutation in a HOX gene that caused this reversal, that has had such interesting developmental consequences, including endoskeletons and exoskeletons, respectively. Each kind of animal was set off on very different developmental pathways due to this one strange bifurcation in developmental orientation.

261.

Arthropods in Reverse

The arthropods are reverse vertebrates.
The insect's mouth was once the mammal's anus,
A reversal which makes for different fates:
An exoskeleton that makes a heinous
Soft crunch is fulcrum to the muscles knit
Inside, the opposite of muscles linked
To bones. Our soft and hard are opposite.
Yet, we're alive in the time they all blinked.
We live together in such different ways –

We sacrifice for others, altruisms
The opposite as well—we want our days,
Rejecting insect suicidalisms.
   Two forms of life have taken over earth,
   Though inverse means resulted in their birth

262.

Where does neoteny fit into this picture? Sponges are one of the first true multicellular organisms. A polycellular organism is one where all the cells stick together. Things such as amoebae live freely. And multicellular organisms? Sponges are made up of different kinds of cells with the same genetic code—with some sticking together while other amoeba-like cells moving around among the cells bring food to the stationary cells furthest away from where the food is absorbed. Blood cells in more advanced animals provide the same service. Thus, multicellular animals are halfway between polycellular organisms, which can only become so complex, and single-celled organisms, which are the "youth" of the polycellular organisms. Plants seem to be an exception to this rule—but the first thing we should notice is that plants can apparently become only so complex. This is likely due to the rigidity of their cell walls, which provides the stability the plants needed to become more complex. But they were not able to achieve the kind of betweenness that allowed for the kind of massive, rapid complexity-creation found in animals. Not everything in the universe reaches new levels of complexity that allow for the creation of new, more complex fractal systems, and not everything in the universe reaches similar levels of complexity at the same time. It is interesting to note, however, that dicot flowering plants, being more organized, proliferated in kind and number over their less organized ancestors, but monocots, which evolved more recently, are less organized in structure (their internal structure could be seen as on the borderland between order and randomness) than are dicots, yet have proliferated even more rapidly, and are even more complex in many ways. Look at the expansion of the grasslands, and the fact that the type of plant with the most species overall—25,000—are orchids, a monocot, and the plants with the most complex reproductive systems. So perhaps plants are beginning to catch up with animals a little.

263.

HOX genes make patterns. However, the fact that they are highly conserved suggests that they have the kind of stability that is necessary for the proliferation of variation—in the same way that carbon's stability makes it perfect for the proliferation of variation in life. HOX may be the element that developed to go from mere poly-celled eukaryotes to true multicellular organisms. The creation of patterns may have been precisely what drove this kind of development.

264.

Cells can be understood as liquid crystals. Crystals are highly stable, but a true liquid is too chaotic. A liquid crystal has highly restricted movements. In the

On the Creation of Complexity in the Universe 101

same way, HOX genes are very stable but, like all other genes, prone to mutation. Most mutations are fatal, but just a few may result in substantial positive changes. We could perhaps think of HOX genes as a sort of quivering hand—they cause things to happen, but their quivering can cause unexpected things to happen. Some of these things will be good, most will be bad. Like an artists with a quivering hand, most mistakes will be just that, and the work will have to be reconsidered (selected against)—but occasionally that "mistake" will be exactly what the work needed. Further, the stability of the HOX genes would also perhaps give other genes more leeway in trying out new mutations (of course, so does redundancy in eukaryotic DNA)—there are different genetic clocks for different kinds of genes.

265.

But let us return to the increasing complexity found in animal evolution. Once HOX genes evolved, there was a proliferation of invertebrate forms. One such form was the sea squirt. The adult sea squirt is reproductive, but immobile. The larval sea squirt is highly mobile (it swims), but is nonreproductive. There exists an in-between creature called the lancelet. It is almost identical to a larval sea squirt: it swims, has a notochord, etc. However, it has an advantage: it can reproduce. Thus, it never has to mature into an immobile adult sea squirt. Instead, it is able to swim around and avoid predators and fill many niches in the ocean other than rock faces. The lancelet is ancestor to the bony fishes and, subsequently, to all the vertebrates. The lancelets are generalists in relation to the sea squirt, and thus have much smoother fitness landscapes—the vertebrates they gave rise to have a very large variety of rugged landscapes as the numbers and kinds of vertebrates proliferated.

266.

Chimpanzees live very complex social lives. They also have a reproductive strategy—one infant every few years—that has pushed chimpanzees and the other apes into a problematic situation that could very easily have resulted in extinction (there are, after all, very few species of ape left). This problem was resolved through the evolution of humans through neoteny, which allowed humans to have offspring more often. But with that advantage came many more. Adult chimpanzees reproduce, but are not very creative; young chimpanzees cannot reproduce, but are very creative—any cultural changes typically start with discoveries by younger chimpanzees. However, adult humans are both reproductive and highly creative. Further, humans have a very smooth fitness landscape. Kauffman points out that recombination occurs most often on flatter fitness landscapes. As it turns out, what separates humans from chimpanzees is not so much percentage of genetic difference (we are around 98% genetically identical), but that we have massive amounts of genetic recombination, including regions of inverted DNA and fused chromosomes. Because we have smooth fitness landscapes, we can live in any habitat. Our ability to live in any habitat evolved first in *Homo erectus*, which spread from Africa across Europe and Asia. It seems that at this point population dynamics overtook natural selection. Only later did *Homo sapiens* evolve, with our more complex brains, which pushed our brains into a far-from-equilibrium state, and broke the symmetry created by the evolution of *Homo erectus*. Natural selection took over again long enough for *Homo sapiens* to become the only hominid in the world, after which population dynamics again took over. We could do this because, with the emergence of complex thought, humans became biologically stable. When we developed language, which likely came about due to a bifurcation of a mating call into language and music (which is perhaps why we feel such a deep connection to poetry and song), our creativity blossomed into our cultures, arts and literature, and technology. Humans evolved into a kind of habitat symmetry with minds on the edge of chaos, which resulted in the creation of a smooth fitness landscape for culture, art and literature, and technology. This is why free cultures and economies are so creative.

267.

I have outlined the specifics; now I will explain the general mechanism, or physical law(s), which underlies all of these specifics. What I am proposing is the mechanism for the creation of more objects, and more complex objects and emergent orders of complexity (nested hierarchical systems): symmetry exists at far-from-equilibrium conditions; far-from-equilibrium conditions are generative of new objects, which break the symmetry as they organize themselves; as more objects accumulate, there is a search for the smoothest fitness landscapes, or new symmetry. Each new symmetry breaks in such a way that the original symmetry holds, while a new fitness landscape develops that is asymmetrical. Evolution occurs such that the fitness landscapes evolve toward increasing smoothness for at least one system. Once smoothness, or a new symmetry, is reached, a new set of fitness landscapes emerge with the emergence of the new, more complex level from the far-from-equilibrium state. Both a crystal and a gas in a container at a constant pressure and temperature are at equilibrium—but crystals and gases are asymmetrical. It is symmetry which appears to in itself put a system in a far-from-equilibrium state by keeping the system on the edge of order and chaos—precisely where it needs to be in order to be creative and generate new things. So symmetry and equilibrium are opposite things. To get the kind of symmetry needed to get this kind of emergence into a new, more complex level, a system must take a sort of half-step back (into neoteny) so a smoother fitness landscape can be created, before there is the leap into the next level of complexity, where symmetry is broken. For each complex system, there is a nested hierarchy of systems with smooth fitness landscapes. Complexity-creation requires that one first become a generalist. Thus, the smooth fitness landscape of human culture emerges from the smooth fitness landscape of our being vertebrates, which is built on the smooth fitness landscape of our being eukaryotes, which is built on the smooth fitness landscape of our being life, which is built on the smooth fitness landscape of organic chemistry, which is built on the smooth fitness landscape of quantum physics. Each of these levels of course have systems in them with rugged fitness landscapes—but to get increasing complexity, one must have a hierarchy of smooth landscapes. Thus, all the other vertebrates have more rugged fitness landscapes than do humans, who have emerged a new level of mental complexity with various fitness landscapes for each culture, and perhaps for each person (making each person a strange attractor). This is likely the source of the proliferation of cultural and technological objects.

268.

Stuart Kauffman is correct to suggest we are missing something. We have failed to see that symmetry is the opposite of equilibrium, meaning to reach equilibrium, we have to break symmetry, which creates more things in the universe, keeping the universe far-from-symmetry. I hope, in showing that more things are generated in the universe because symmetry puts a system in a far-from-equilibrium state that causes the self-organization of new things that break the symmetry while some of the things that broke the symmetry search for a new symmetry, that I have at least suggested what we have until now failed to see. The universe itself is attracted toward both symmetry and equilibrium. As such, it cannot help but be generative of more and new objects. This is the source, not only of all of the objects in the universe, and of the arrow of time, but all of the choices in the universe—meaning, that the universe is inherently free.

A Fractal Model for Emergence in the Universe to the Metahuman

269.
A radical proposal: a new form of complexity has emerged from the human: the metahuman. If the theory of emergence is true, then just as molecules are the bridge from quantum physics to life, then the human is the bridge from the animal to the metahuman. For those who expected whatever was to emerge from humans to be radically different from the human, there is bound to be great disappointment; but for those who understand how fuzzy the differences are between complex cycles of carbon chemistry and the simplest of prokaryotic cells, or between chimpanzees and humans in our behaviors, then it will not be too much of a surprise to learn that the metahumans among us will be very similar and familiar to the average person in behavior, even if metahuman thinking, on closer inspection, is quite different, and even strange to the average human being. The metahuman is not physically different from the human—he or she is mentally different, emergent from human thinking, and nonetheless inclusive of all its forms.

270.
Emergence occurs when the parts of a system interact with each other in ways so complex that the system as a whole develops characteristics that cannot be predicted from its parts. The whole is greater than the sum of its parts. A living cell is a good example of such a phenomenon. In *At Home in the Universe* Stuart Kauffman defines life as "an emergent phenomenon arising as the molecular diversity of a prebiotic chemical system increased beyond a threshold of complexity." This life "is not to be located in its parts, but in the collective emergent properties of the whole they create" (24). But complexity and emergence are found at more levels of the universe than just life. We see quantum physics emerging out of the pure energy of spacetime. We see chemistry emerging out of quantum physics. And we see human intelligence emerging out of life. And each of these new levels emerge as a complexity threshold is surpassed.

271.
Again, the word "complex" means "folded," and this idea of folding helps us to understand complexity. When we get a quantum physical particle-wave emerging from the spacetime energy, we get it from spacetime folding. In quantum physics, when a wave breaks up into trillions of wavelets to give emergence to the particle form, what we have is a folding of spacetime—and this folding causes spacetime to give rise to new qualities. When these particle-waves interact with each other to create more complex atoms, we have spacetime fold-

ing in on itself even more. There is increasing folding as atoms interact to create molecules, molecules interact in complex cycles to create living cells, cells inter-act to give rise to organisms, organisms interact socially, and neurons interact to give rise to complex thought. Thus, we are not *in* spacetime, as though we were in a container, but we are spacetime, only a more deeply folded region of it. Now, if we think about the qualities of fractal geometry—a deeply folded border giving rise to a finite space increasingly surrounded by a border approaching infinity, resulting in self-similarity regardless of scale—we can see that a human being is fractal at a higher scale than is any other living organism, which is at a higher scale than molecules, etc. The fact that humans are at such a high level of spacetime folding has several consequences, not the least of which is our very complex behavior.

272.

To look at complexity from another angle, we see that to have complexity, we must have a situation that is "orderly enough to ensure stability, yet full of flexibility and surprise" (Kauffman, *AHU*, 87)—in other words, it is near the phase transition between order and chaos. If we have, say, a molecule that is too orderly, such as a sodium chloride crystal, then nothing interesting happens. We can predict with 100% accuracy what will come next in the sequence: Na-Cl-Na-Cl-Na-Cl- . . . , so long as we have more crystal. At the other extreme, with pure randomness, such as we have with a gas in a container at equilibrium, we cannot predict at all what will happen. This is as uninteresting and uncreative as pure order. But in a region of phase transition, in an edge-of-chaos regime, we have complex interactions, swirls and eddies, a combination of the predictable and the unpredictable. In other words, it is like a good story, which can be neither purely ordered and predictable nor disordered and unpredictable, but must have elements of both in order to be enjoyed. We should not be surprised that we enjoy (which is to say, our brains reward themselves for recognizing) things similar to ourselves, for doing so has an evolutionary advantage. Any species or individual who is better able to recognize the fractal geometry of the world would have an evolutionary advantage over those less able to do so in a living world of fractal geometry.

273.

J. T. Fraser proposes a model of emergent reality that ties into this idea of emergent complexity, while explaining the evolution of time over time in the universe. From his model, we learn that the experience of time changes over time, and that evolution evolves. In brief, if we start from the Big Bang, the moment of pure energy, pure randomness, we begin with no time experience—as we learn from Einstein's theory of relativity, anything traveling at the speed of light will not experience time passing. As the universe expanded and cooled, quantum physics with its many forms of particle-waves emerged—and with them, a probabilistic time experience. In the wave-form, there is no time experience, but in the particle-form, there is deterministic time experience. As atoms, and then molecules, emerged, deterministic time emerged—this is the

A Fractal Model for Emergence

experience of time Newton described. With the emergence of cells from complex systems chemistry, an experience of time as happening in a particular direction emerged. And with the emergence of human intelligence, there was the emergence of the experience of past, present, and future in a very strong way—since we can think about things that happened before we were born, and things that will happen after we die. The more spacetime folded in on itself, the more time was gained by each level—first, the present was gained and then, with animals, a limited past and future, and then, with humans, even more past and future. One would expect the next level of complexity, having even more folds of spacetime, to have even more time experience.

274.

But even this is only a simplified description of the world. The deterministic level is not purely deterministic, but also contains elements of the probabilistic and the truly random, from the two levels below it. Only the strictest of solid-state physics acts in an almost purely deterministic fashion. But complex systems chemistry makes strong use of probabilistic elements from the quantum physical state. In the same way, with living organisms, there is not only the emergence of freedom in life, but deterministic, probabilistic, and random elements as well. Darwin focused on the random elements in his theory of natural selection, but more recent work have shown that there are probabilistic, deterministic, and even primitive choices in evolution as well. And all of these elements are also present in the emergence of human intelligence, in which emerged even greater freedom in the ability to model many more multiple futures from which to choose. The universe does not get rid of lower levels of complexity, but rather builds on top of them, enfolding and incorporating the lower levels into the new emergent levels. Further, the spacetime field, with each folding into more complexity, becomes increasingly individuated. What we see in the emergence of each new level of complexity is the emergence of even more individuation and even greater freedom—and that is what we would expect in the emergence of the next level of complexity as well.

275.

Objects with fractal geometry have self-similarity regardless of scale. If the universe itself is fractal in its geometry, we would expect self-similarity to be expressed in the emergence of new levels of complexity as well. In other words, there should be two sublevels of complexity at the quantum physical level, since it is the second level of complexity, and three sublevels of complexity at the chemical/macrophysical level, since it is the third level of complexity.

276.

The first level, the pure energy of the spacetime field, has but one level—that of pure energy. It gives rise to the second level of reality, that of quantum physics, and with it two sublevels of reality—that of free particle-waves, such as photons and electrons, and from them, the emergence of much more complex atoms, which are particle-wave systems. From here we get the emergence of chemistry in the reconciliation of the paradoxical need to have both full electron shells and charge neutrality. At this level, we get the emergence of three different sublevels of molecular/physical reality—fluid dynamics, solid-state physics/chemistry, and complex systems chemistry (which emerges on the borderlands between fluid dynamics and solid-state physics/chemistry). From complex systems chemistry, we get the emergence of biology. And since it is the fourth level, there are four sublevels—single-celled life forms (eubacteria, archaebacteria, and eukaryotes), multicellular life forms (such as plants, animals, fungi), and, in the animals, we have the line that led to vertebrates, invertebrates in the line that led to those with exoskeletons, and the predecessors of both, and in the vertebrates, we have schooling/herding, territorial independent, and territorial social animals. I will note only in passing that each of these four subdivisions themselves have three subdivisions.

277.

The next level of complexity is human intelligence –Fraser's nootemporality. In this level have emerged five levels of human mental complexity, as outlined by Clare Graves, Don Beck and Christopher Cowan. Actually, they came up with six levels, but the first level of mental complexity is actually that of social mammals, particularly the social apes, as ethological research showa. Thus, there are in fact only five levels of emergent mental complexity in humans, into which I will go into more detail later. I will only point out that the last sublevel of mental complexity is what prepared the groundwork for emergence into the next level of complexity, what Graves, Beck and Cowan have called Second Tier thinking, and what I am calling a new level of complexity on the same level as the emergence in complexity from the animals to humans.

278.

Clare Graves, Don Beck and Christopher Cowan recognize six levels of emergence in human thinking. Again, I recognize five, since the first one is actually the same level of thinking as chimpanzees and bonobos (possibly even social mammals in general), though it also belongs to humans. Beyond that first level, which belongs in the biotemporal level, there are five distinctly human levels of emergently complex thinking. The first is tribal/familial thinking. The second is the heroic/egocentric level, when writing is invented in a culture, when we hear tales of people typical of this level, such as Achilles, Odysseus, and Gilgamesh. The third is the level of truth from authority, typified by such thinking as that of the medieval Christian church and social conservatives. The fourth is the level of capitalist and deterministic scientific thinking, typified by materialism and liberal/libertarian thinking. The fifth is the level that promotes human bonds, and typically is anti-hierarchical, egalitarian, relativistic, perspectivist, and typically respectful of differences among people (at its best; at its worst it is dehumanizing, politically correct, fascistic, and Marxist). This fifth one is a smooth fitness landscape, of flattened hierarchies, and thus is the groundwork for emergence into the next level of complexity.

A Fractal Model for Emergence

279.

Just like with the biological level of humans—where we are social mammals, vertebrates, multicellular, and cellular simultaneously, as well as chemical and quantum physical and ultimately made up of energy—emergence into each new level does not mean elimination of the lower levels. Rather, we have a nested hierarchy. If we take someone who is at the top level of human thinking—the egalitarian level—we see that they can contain all of the lower levels as well. In addition to promoting human bonds and having a multidisciplinary view of the world, they typically believe in science as a way of knowing the world and are disciplinary, they still need ethics and believe in categories, they still have mythic needs, and they also have the need to belong to a family and have rituals. The difference is, each of these levels tends to be more inclusive of other people. The first level is that of the family and tribe—and thus quite exclusive. The second level expands into wider cultural contexts, such as being Greek, being Egyptian, etc. The third level expands into a wider group of fellow believers, regardless of culture or language, such as being Christian, being Buddhist, etc. The fourth level expands into including anyone who will trade with us or will engage in the pursuit of scientific knowledge with us. The fifth level includes all people in the world—but especially those who are of the same level (there is a bit of a tendency at this level to suggest that all animals are equal, but that some animals are more equal than others—though, officially, everyone is in fact the same). But in each, the family (lowest level of thinking) remains—even if our ideas about the family change in each new level. In the same way that biology affects the chemistry that gives rise to it in a bidirectional feedback loop, each new level of thinking affects lower levels of thinking that gave rise to it. The family for a fifth level person is different than is the family for a second level person.

280.

Each of the physical levels of existence are also environments. Humans create a cultural environment humans have to live in, though if this environment were to disappear, it would harm nothing on earth except humans (which is actually to get it backwards since, so long as there are humans, there will be human culture, so it takes the destruction of humans to get rid of human culture). Plants, fungi, prokaryotes and animals, including humans, all create and affect the biosphere and all the various regional bio-environments. Humans are imbedded in this environment as well, and cannot survive without it, being biological organisms. Life and the biosphere are in intricate, intimate interconnection. But this bio-environment is also embedded in a larger chemical environment. There are geological processes going on (many of which are affected by the biosphere in a nonlinear feedback loop), atmospheric chemistry and weather, water and hydrophilic chemistry, etc., which create the baseline on which life emerges and engages in its own kind of chemistry. Without this environment, there would be no biological nor cultural environment. Next, the chemical environment is imbedded in a quantum physical environment of particle-waves, which creates the background and baseline on which chemical and macrophysical processes occur.

And quantum physics emerges out of an environment of pure energy. None of the new, emergent environments could exist without the environments below them—thus, a sort of environmental pyramid emerges. If any of the levels are removed or damaged, everything above that level collapses with it. Thus, humans have to learn to live better in each of these environments, and to see themselves as living in these environments. Further, each of the levels of human thinking create their own cultural environments—which we have to learn to live in even as we emerge into new levels of thinking. However, it is the top levels that are obliged to live in the environments of the lower levels, not vice versa, as the lower levels can and will continue to survive even if more complex ways of thinking emerge. Attempts to eliminate lower levels will result in tragedy.

281.

All five emergent levels of human thinking also reflect the time experiences of all the levels, below and including the human. Time is experienced by pure energy as atemporal, meaning the time experience is circular. Incidentally, the tribal way of thinking experiences time as circular in an eternal return of the same things. The second physical level of probabilistic time, going between circular and linear time, is also shared by the second level of human thinking as an eternal return of similar things (circular time linearized into a spiral or helix). The third physical level of chemistry/macrophysics' experience of time as deterministic is shared by the third level of human thinking, where human lives and history are understood to be on a certain path, often one already prepared by God. The fourth physical level of biology, which experiences a slight forward direction of time, is also shared by the fourth level of human thinking, since for the first time life and history are understood to change and be changeable. Finally, the fifth physical level—the human level—is doubly experienced at the fifth level of human thinking, where past, present, and future are beginning to be brought together into a single model (which sometimes has the ironic outcome of giving rise to ideas of an "unchanging state of nature" that does not exist and never has and never will). And, just like in the physical world, the human experience of each of these levels bleed through into higher levels, creating even more complexity than these apparently simple divisions indicate. But do not forget: models, like math, are only precise approximations of reality. Models are digital representations of a digital-analog world.

282.

With this model we see some of the parallels between the human levels of thinking and their emergent complexity, and the emergent levels of physical reality. Further, we have seen that each new emergent level has become more internalized, and individualized, ending with the emergence of one particular species into complex intelligence. One would expect, if new levels of emergent complexity emerge as past levels have, with each emergence becoming even more individualized, that individual humans would be the ones emerging into the next level of complexity—Graves, Beck and Cowan's Second Tier thinking.

A Fractal Model for Emergence

And this leads us into my proposal that this Second Tier thinking is in fact the emergence of a new level of complexity.

283.

J.T. Fraser says the next level of emergence is the sociotemporal level, which he says emerges from the interaction of humans in society. But this makes as much sense as saying the level emergent from the biotemporal is the world ecology level of time experience. The social actually evolved prior to humans, in the social mammals, and the way that Fraser uses "sociotemporal," it is clear that he means something like the world culture. But as we have already seen, the culture is merely the environment in which the nootemporal lives. The nootemporal level, the actual next level of complexity out of the biotemporal, evolved from a particular kind of animal, and the next level of complexity will evolve, not from all humans, as the idea of sociotemporality suggests, but from particular humans—those who, after reaching the highest level of human thinking, have seen and resolved the paradoxes of human thinking, and have thus emerged into the next level of complexity. In fact, I would argue that it already has evolved in Graves, Beck, and Cowan's Second Tier thinkers. The first two sublevels within this new level of complexity are integrationism and holism. Others will follow and, if the same pattern that has been found in the universe throughout its entire history holds, it will have six of these sublevels before the paradoxes inherent within this level create the conditions for emergence into the next level. And, just like humans have all the other levels below them, including the biotemporal level, making humans in that sense indistinguishable from animals, the Second Tier thinkers are indistinguishable from other humans, except in certain aspects of their thinking. I would propose calling this new level not sociotemporal, but intertemporal, since it is the first level self-aware in understanding both the depth of its own thinking, through all the sublevels of the nootemporal, as well as the nested hierarchical temporality of the universe as a whole. In this sense, the intertemporal level is also neotenous, in that it takes on the fully "adult" way of human thinking, but also every other level under it, as well as embracing every other level of reality, and not just the "adult" one of nootemporality.

284.

Beck and Cowan say that Second Tier thinkers have several qualities. While human thinking has its fears and each level knows that it has the answer, in metahuman thinkers, fear drops away, and what is known is precisely how much is not known—such thinkers know so much, they realize how much is inherently unknown, and even unknowable. This emergent level is where emergence becomes understood and known. This is where it is most self-reflexive; it is the first level of emergence that comes to be fully conscious of emergence as such. And it becomes aware of its own self-emergence from human thinking, whether it has found the words to articulate the idea yet or not. Here too there is the first real awareness of time—no effort is made to eliminate time from consideration, as human thinking tries to do in notions of eternity, timelessness, and "unchanging nature." Further, Beck and Cowan point out that the Second Tier person "sees too much, from too many new angles to accept simplicity that is not there" (273). There is an inherent interdisciplinarity in metahuman thinking. Interdisciplinarity is different from postmodern (egalitarian human) multidisciplinary thinking; postmodern multidisciplinary thinking is pluralist, poststrustructuralist, and anti-hierarchical, making it tend to make false connections and incorrect associations since everything is considered to be on the same level and is therefore fundamentally the same. Postmodern thinking tends to be deeply reductionist (like deconstruction) and ahistorical (since, if nothing has changed, why bother with history?). For it, everything has the same level of complexity—meaning, if we understand quantum physics, we will understand the complexities of biology, human thought, and culture. As opposed to multidisciplinary thinking, interdisciplinary thinking is based on fluid and nested hierarchies which place everything in proper relation to everything else. Quantum physics is at a level of complexity below chemistry. Humans are not divided up according to race, but according to complexities of thinking, which occur based not on race, but on life conditions. Interdisciplinary thinking is emergentist and complex, as well as historical and evolutionary. It is information- and knowledge-driven. It understands the world is nonlinear, meaning top levels affect lower levels once the top levels emerge, but the lower levels are what give rise to the top levels. Thus, there is both bottom-up determinism (which is foundational), and top-down determinism (which emerges), both working in a nonlinear, and thus chaotic or biotic, fashion.

285.

There is an exponential (sigmoidal, actually) difference between linear, human thinking, and nonlinear, metahuman thinking. "With the shift toward Second Tier thinking the conceptual space of human beings is greater than the sum of all the previous levels combined with a 'logarithmic' (Graves' term) increase in degrees of behavioral freedom" (Beck and Cowan, 276). Although Beck and Cowan then go on to deny that this is a "new breed of human" (and in a real sense, they are correct that it is not a "new breed" of human per se—as this new level of complexity cannot be bred for), this is precisely the description of emergence into a new level of complexity. Each new level is logarithmically more complex than the previous level. In *Time, Conflict, and Human Values*, J.T. Fraser proposes that there have been 10^{1000} organisms through the history of life on earth (he also suggests we would get a complexity of 10 at the quantum level, and 10^{10} for the level of chemistry/macrophysics), while for humans, we would get a level of complexity of about $10^{10,000}$, for the number of possible brain states. This then suggests that the next level of thinking would be at a level of complexity equal to $10^{100,000}$. If the brain is emergent in complexity from life itself, then the brain should give rise to the next level of emergence in its own complexity. Thus internalized, this new level of complexity would be very difficult to detect—and could not be detected by just looking at different people—but only by seeing how certain people think and behave.

A Fractal Model for Emergence 113

286.

What is metahuman thinking like? Unlike at human levels, there is an understanding for such people of the legitimacy of all levels of thinking and existence, as well as an understanding of these levels' importance and proper position in the world's natural hierarchies. They have a sort of extreme self-awareness that is accompanied by self-acceptance. And they are adept at integrating complexity, explaining parallels, and creating and seeing connections among things. Such thinking is thus at least highly interdisciplinary, and even holistic in nature. Ideas are multidimensional, paradox and uncertainty are not just seen, but understood, and even enjoyed. Chaos and order are understood in their proper relation to each other—not as in annihilating oppositions, but in creative conflict. Multiple perspectives are considered simultaneously, given proper weight, and used to inform each other before a decision is made. Thus, difference is of utmost importance, though the mistake of thinking that difference necessarily means either good or bad is not made, as it is understood that difference just means difference. The world is understood to constitute particles and entities as well as groups, fields, and waves, and it is understood that there is a "'holistic' wisdom within systems" (Beck and Cowan, 284), and that this wisdom is not in conflict with knowledge—that the two together in fact constitute beauty. The world consists of fractals, the laws of nature that apply throughout the universe supplant doctrinaire laws, and it is understood that everything connects to everything else. Fortunately, with the higher level complexity thinking involved, all of this can be done extremely rapidly.

287.

What is perhaps most interesting is the issue of communication of information. With each new level of complexity, there is the emergence of new forms of communication. Quantum physical bodies communicate with each other using particle-waves—electrons communicate using photons, for example. Molecules communicate with other molecules using both quantum physical elements, but also topology. Living organisms use both of these, plus molecules and, for some, sound to communicate. Humans use all of these, plus grammatical language—which is to say, combining sound communication with the narrative structure required to make active decisions and choices. And what of metahuman communication? First, it appears that metahuman thinkers ar able to think while communicating. But there is more. In reading the next part, one must keep in mind that, to other metahuman thinkers, this will make sense, while to human thinkers, it will make as much sense as human language makes to other animals. This is not an insult—it is the nature of emergence into new levels of complexity. And the fact that I have to communicate this in language only makes it more difficult, as it is not the proper form in which to communicate this kind of information. But please bear with me on this, as what will appear as borderline insanity to many will make complete sense to some. But this is no more or less insane than building a space station is insane to a cat—a space station makes sense to the human way of thinking, but to a cat, if you can't eat it, drink it, or have sex with it, then to do something like build a space station is at best nonsensical; at worst, madness. Beck and Cowan in fact argue that 2nd Tier thinkers are practically invisible to 1st Tier thinkers, just as more complex 1st Tier thinkers are practically invisible to the least complex 1st Tier thinkers. The less complex thinkers just can't see what the more complex thinkers are thinking.

288.

Frederick Turner, in *The Culture of Hope*, talks about some things that, at first glance, from a human way of thinking, appear borderline mad. He talks about how the poet is able to communicate with trees, stones, mountains, etc. What many would take to be a poetic metaphor, Turner means in a literal sense. What I once took as poetic I now understand much more clearly and concretely: the metahuman thinker is able to communicate with trees, stones, mountains, etc., which do not communicate using human language, but in their own languages. The metahuman thinker is the first to be open to such communication, and is the first to be able to understand across the levels of complexity. This can be understood if we really understand what is meant by spacetime being more folded with each new level of complexity. Spacetime is in contact with itself more and more. Thus, a more complex level than humans should be able to communicate even more, and more clearly, with and through more levels of spacetime. This leads us to what appears to be an even stranger form of communication at the metahuman level. With human language, there is a limited reflexivity—the present refers to a close past in order to push into the future. This gives humans a great deal of freedom, and an ability to greatly order the world. Each new, emergent form of communication has increased in reflexivity, and given more freedom, and more order, to the world. Thus, we should expect the new, emergent form of metahuman communication to fit that criteria as well. And it does. Beck and Cowan talk about how Second Tier thinkers are more intuitive—such thinkers know what they should be doing at any given time, especially when dealing with big decisions. Frederick Turner suggests that these intuitions are people in the future communicating backwards in time to the present to guide us. This makes sense if we realize that communication for the metahuman has become even more reflexive, so that all potential I's in the future are able to reference my past I (which is actually my present I) to guide me into the best path(s) to create a better future. Perhaps even more complex future levels are more capable of communicating through time this way, having even more spacetime folds in them—and this new level of complexity is the first to most clearly receive this form of communication, having now been deeply folded enough into spacetime to do so. The way to understand this is to see the future as branching, and all our future selves as well as others' future selves on all the branches. Some of the branches are better futures than others. For the metahuman, the future I's (perhaps not just theirs, but others') are able to communicate back to let the present I know what would be the best path. If there is a bad future, that particular I would either be silent, or discourage such a path as would lead to that future, while for good futures, those I's would encourage decisions that led to them. Spacetime folds back onto itself in greater complexity than we now understand, and thus, because we are able to communicate with ourselves backwards through time, sending these "feelings" or "intuitions" back to let us know which paths are best to take, we have both greater freedom and more order in the universe. Now, all of this would be confusing to those who first encountered such communication—the same way, I would imagine, those who first emerged into grammatical language found it confusing and disconcerting though, as they grew used to it, and played with it, and learned how it worked, it turned into a great and powerful tool that molded not only their thinking, but future human generations' thinking. As the metahuman level comes to understand this new form of communication, and how to use it better and more efficiently, they will grow more comfortable with it, and more able to communicate with it.

289.

If I am correct, that some humans' thinking has emerged into a new level of complexity in the universe, then much more work needs to be done to learn more about this new level. Of course, this research can only be done by those who have emerged into the metahuman—levels can only understand their own levels, and those underneath them. But at the same time, the metahumans need to be aware that that is what they are. Thus, this chapter is in fact addressed to those people who are already metahumans—and perhaps to those who are on that borderland of egalitarian order and chaos that will give rise to metahuman thinking. Who knows how many people are in this latter category, or what will push them into the next, emergent level? Nature solves the paradoxes that arise in each level of reality by emerging into a new level of greater complexity—which creates its own, new paradoxes. When the paradoxes of human thought become too much for the last level of human thinking in such people, then they will emerge into the next level of complexity. As for the human thinkers, all of this will seem too subjective, too based on private experiences that cannot be tested in any sort of objective, scientific manner. To some, it will even seem like madness. To that, I would suggest that we should not so quickly disregard private experiences—we should not deprive ourselves of any of the multiple perspectives that make up the world, even if we do end up partial to particular ones. Judge us, rather, by our actions. If what appears to be madness gives rise to good works in the world, then that is the kind of madness the world needs.

290.

Like anything dealing with human behavior and thought, an idea like this one can and likely will have political consequences. The first thing to note is that there is not even the remotest racial element to this: every race can and does have individuals at each level of human thinking, and each and every race will give rise to metahuman thinkers as well—as genetic research has shown, humans of all races are practically genetic clones of each other, meaning there is no genetic reason why anyone in any group cannot emerge into new levels of thinking. Women as well as men can also emerge into this new level. Further, metahuman thinkers cannot be bred any more than different levels of human thinking can be bred for. Metahuman thinking emerges out of human thinking, which is affected by psychosocial life conditions. This is the environment for mental development in the healthy human being. It is possible for social conservative thinkers to have metahuman offspring, if those offspring are raised in a complex enough society. So race and eugenics are useless, as is a purposeful attempt to create metahuman thinkers through education, nurture, or other societal interventions. Certainly education, nurture, and society (as well as genetics—though the genetics for behavior are at best only terribly understood) all contribute to the emergence of metahuman thinking—but one person raised in the same conditions will only achieve certain levels of human thinking (and not necessarily the most complex level), while another will emerge into metahuman thinking. Some will do it early in their lives, others later—though I suspect that, like emergence into human thinking—it must be reached by a certain stage, or it will never be reached.

291.

Animal rights activists are right about one thing: just because humans are more complex, that does not give us license to be cruel to animals. In fact, we should try to show greater compassion—to animals as well as to each other. The same should also be true with the relationship between metahuman and human thinkers. In fact, I would argue that metahuman thinkers would be even more compassionate toward other levels. Perhaps metahuman thinkers will become less political—in the human sense of the term. Less political in the power sense of the political. But, in the sense of caring about the well-being of the *polis*, I also believe metahuman thinkers will become more political. With metahumans, all human needs will remain—but not necessarily all human wants. Beck and Cowan suggest that Second Tier thinkers tend to become more minimalist in their physical wants—though their mental wants and needs do increase exponentially. Metahuman thinkers do not want power—but they do feel the need to help their fellow man. There are no dictators among metahuman thinkers—only federalist democratic republics have the complexity needed for the political to map well onto nature, including human nature, and metahuman thinkers know this. Further, metahuman thinkers, being in a sense on the outside, are able to see the systems emergent from human thinkers, including political, economic, and social systems, as well as cultural products. But they are also aware of how complex the world is, of how much they do not know or could possibly know, and so do not think it possible to create utopias, which always replace freedom and creative order with oppressive order. The political consequence of the emergence of metahuman thinking will be even greater support for political, economic, and social systems structured to be on the border or order and chaos—which is to say, freer, more ordered, and, as a consequence of both, more creative.

292

In "The Self-Organizing Quantum Universe," Jan Ambjørn, Jerzy Jurkiewicz and Renate Loll demonstrate how four-dimensional spacetime can emerge from

A Fractal Model for Emergence

the self-organization of fractal spacetime elements, so long as time is considered real and oriented in a particular direction (just like we experience it). They argue that, like a snowdrift, space is made of fractal elements that, on a larger scale, smooth out and become 3-D space (or 4-D spacetime, once oriented properly using time). When time is oriented consistently among the elements, cause and effect become unambiguously distinguished, menaing "the distinction between cause and effect is fundamental to nature rather than a derived property" (*Scientific American*, July 2008, 46). So it seems very likely that the universe at its most basic level is fractal, renormalizable, self-organizing, dynamic, time-dependent, and emerg-ent. We should not be surprised to find these features at each new level of complexity in such a universe.

293

The universe is an open system. If it is an open system, it is in a steady state (which does not go against the idea of the universe expanding, growing, and changing). Ludwig von Bertalanffy observes that open systems, by "maintaining themselves in a steady state, can avoid the increase in entropy, and may even develop towards states of increased order and organization" (*General Systems Theory*, 41). Since we know the universe has developed "towards states of increased order and organization," that means it must be an open system maintaining itself in a steady state.

294

There are no natural systems controlled by a central authority.

295

A natural economy is a bottom-up hierarchical self-organizing open system. A natural society is a bottom-up hierarchical self-organizing open system. A natural culture is a bottom-up hierarchical self-organizing open system. Bottom-up hierarchical self-organizing systems contain a great deal of redundancy, which is what makes them robust. The more complex a system is, the more redundancy it has to have to remain stable. Remove redundancy—that is, simplify the system and make it more "efficient"—and you kill the system. If one part of a completely efficient system breaks down, the entire system breaks down. This is why socialism is always going to be a failure—as though it being a dehumanizing system were not enough of a reason. Systems theory warns us "that the Leviathan of organization must not swallow the individual without sealing its own inevitable doom" (Bertalanffy, 53).

296

Most of the errors made within the disciplines are due to the disregard for time. For example, "history is sociology in the making or in "longitudinal" study. It is the same sociocultural entities which sociology investigates in their present state and history in their becoming" (Bertalanffy, 8). Perhaps sociology suffers the effects of anarchy and therefore needs history and anthropology as foundations.

297

In economics, supply and demand curves have a point called the equilibrium point. This point only exists in a land without time. What we need is steady-state economics, recognizing the economy as a dynamic system. Still, the supply-demand curve is a useful fiction that more people need to be more familiar with so fewer demagogues will have the ability to successfully lie to us about the causes of prices (competition among producers drive prices down; competition among consumers drive prices up).

298

Human beings are not Economic Man or Social Man or Political Man or Rational Man, etc. They are each of these simultaneously, plus Individual Man, Biological Man, Sexual Man, Religious Man, Evolving Man, Family Man, Psychological Man, Philosophical Man, Instinctual Man, Educatible Man, and Emerging Man.

299

We need diaphysical education. That means not just scientific information, but a full systems education, which includes meaning and values, particularly ethical values. Education is for the full development of each individual's personality toward excellence and greater complexity.

300

Diaphysics is a theory of the universe as beautiful.

Bibliography and Suggested Readings

Adams, Fred. *The Origins of Existence*. New York: The Free Press. 2002

Ambjørn, Jan, Jerzy Jurkiewicz and Renate Loll. "The Self-Organizing Quantum Universe" *Scientific American*, July 2008. Vol. 299 Number 1. Pg. 42-49

Argyros, Alexander. *A Blessed Rage For Order*. Ann Arbor: University of Michigan. 1994

———. "Tragedy and Chaos" *Biopoetics*. Ed. Brett Cooke and Frederick Turner. Lexington, KY: ICUS, 1999. 335-346

Aristotle. *Physics Books I and II*. (Trans. William Charlton) Oxford: Clarendon Press. 1992

———. *On Rhetoric*. Trans. George Kennedy. Oxford University Press 1991
Barnsley, Michael. *Fractals Everywhere*. Boston: Academic Press. 1988

Beck, Don and Christopher Cowan. *Spiral Dynamics*. Malden, MA: Blackwell Publishing. 1996

Bekenstein, Jacob D. "Information in the Holographic Universe" *Scientific American*, Aug. 2003

Bergson, Henri. *Creative Evolution*. Arthur Mithcell, tr. Mineola, NY: Dover Publications, Inc. 1998

Bertalanffy, Ludwig von. *General System Theory*. New York: George Braziller. 1968

Bird, Richard J. *Chaos and Life*. New York: Columbia University Press. 2003

Blake, William. *The Selected Poems of William Blake*. The Wordsworth Poetry Library. 1994

Bohm, David. *Wholeness and the Implicate Order*. New York: Routledge. 1980

Bonner, John T. *The Evolution of Culture in Animals*. Princeton: Princeton University Press 1980

Boole, George. *The Laws of Thought*. New York: Dover Publications. 1958

Borges, Jorge Luis. *Ficciones*. New York: Grove Press. 1962

Boulding, Kenneth E. *The World as a Total System*. Beverly Hills: Sage Publications. 1985

Bibliography

Brier, Søren. "The Cybersemiotic Model of Communication: An Evolutionary View on the Threshold between Semiosis and Informational Exchange" *tripleC* 1(1) 2003 <http://tripleC.uti.at>

Briggs, John. *Fractals: The Patterns of Chaos*. New York: Simon & Schuster. 1992

Burgin, Mark. "Information: Problems, Paradoxes, and Solutions" *tripleC* 1(1) 2003 <http://tripleC.uti.at>

Campbell, Jeremy. *Grammatical Man*. New York: Simon & Schuster. 1982

Casti, John. *Complexification*. New York: HarperCollins Publishers. 1994

———. *Searching for Certainty*. New York: William Morrow & Co. 1990.

Chaisson, Eric. *Epic of Evolution*. New York: Columbia University Press. 2006

Chardin, Pierre Teilhard de. *Christianity and Evolution*. Rene Hague, tr. San Diego, New York, London: A Harvest Book. 1969

———. *The Divine Milieu*. New York: Perennial Classics. 2001

———. *The Future of Man*. Norman Denny, tr. New York: Harper & Row. 1964

———. *The Phenomenon of Man*. Bernard Wall, tr. New York: Harper & Row. 1959

———. *Writings Selected*. Maryknoll, NY: Orbis Books. 1999

Christian, David. *Maps of Time: An Introduction to Big History*. Berkeley: University of California Press. 2005

Clark, Michael. *Paradoxes from a to z*. New York: Routledge. 2002

Coveney, Peter and Roger Highfield. *Frontiers of Complexity*. New York: Fawcett Columbine. 1995

Cox, Christoph. *Nietzsche: Naturalism and Interpretation*. Berkeley: University of California Press. 1999

Darwin, Charles. *The Expression of the Emotions in Man and Animals* Chicago: The University of Chicago Press. 1965

———. *The Descent of Man* New York: Prometheus Books. 1998

Dauer, D.W. "Nietzsche and the Concept of Time" *The Study of Time II* (J.T. Fraser and N. Lawrence, ed.). New York: Springer-Verlag. 1975

David, Michael. *Ancient Tragedy and the Origins of Modern Science*. Carbondale and Edwardsville: Southern Illinois Press. 1988

Bibliography

Davies, Paul. *About Time*. New York: Simon & Schuster. 1995

Davis, Morton D. *Game Theory: A Nontechnical Introduction*. Mineoloa, NY: Dover Publications. 1983

Dennett, Daniel C. *Freedom Evolves*. New York: Viking. 2003

DePryck, Koen. *Knowledge, Evolution, and Paradox*. New York: State University of New York Press. 1993

Devaney, Robert L. and Linda Keen, ed. *Chaos and Fractals: The Mathematics Behind the Computer Graphics*. Providence, RI: American Mathematical Society. 1989

Doczi, György. *The Power of Limits*. Boston & London: Shambhala. 1994

Dozier, Jr., Rush W. *Codes of Evolution*. New York: Crown Publishers. 1992

Dyson, George B. *Darwin Among the Machines*. Reading, MA: Addison-Wesley Publishing Co. 1997

Eisendrath, Craig. *At War With Time*. New York: Helios Press. 2003

Eliade, Mircea. *The Myth of the Eternal Return*. Willard R. Trask, tr. Princeton, NJ: Princeton University Press. 1954

Emerson, Ralph Waldo. *Emerson's Essays*. New York: Perennial Library. 1951

Fischer, Ernst Peter. *Beauty and the Beast: The Aesthetic Moment in Science*. New York and London: Plenum Trade. 1999

Fraser, J. T. *Time, Conflict, and Human Values*. Urbana & Chicago: University of Illinois Press. 1999

———. *Time: the Familiar Stranger*. Redmond: Tempus Books of Microsoft Press. 1987

———. "From Chaos to Conflict" *Time, Order, Chaos*. (Fraser, J. T., Marlene P. Soulsby, and Alexander Argyros, ed.) Madison, CT: International Universities Press, Inc. 1998

Frege, Gottlob. *The Foundations of Arithmetic*. J. L. Austin, tr. Evanston, IL: Northwest University Press. 1980

Fuchs, Christian. "Co-Operation and Self-Organization" *tripleC* 1(1) 2003 <http://tripleC.uti.at>

Fudenberg, Drew and David K. Levine. *The Theory of Learning in Games*. Cambridge, MA: The MIT Press. 1998

Gefter, Amanda. "Throwing Einstein for a Loop" *Scientific American* Dec. 2002

Gell-Mann, Murray. *The Quark and the Jaguar.* New York: Henry Holt & Co. 1994

Gillespie, Michael Patrick. *The Aesthetics of Chaos.* Gainesville: University of Florida Press. 2003

Gladwell, Malcolm. *The Tipping Point.* New York: Back Bay Books. 2002

Gleick, James. *Chaos: The Making of a New Science.* New York: Penguin USA. 1988

Gleick, James and Eliot Porter. *Nature's Chaos.* Boston: Little, Brown & Co. 1990

Goodwin, Brian. *How the Leopard Changed Its Spots.* London: Phoenix. 1994

Gribbin, John. *Deep Simplicity.* New York: Random House. 2004

Gribbin, John & Jeremy Cherfas. *The Monkey Puzzle.* McGraw-Hill. 1982

Gruden, Robert. *Time and the Art of Living.* Boston: Houghton Mifflin Company. 1982

Hall, Nina, ed. *Exploring Chaos.* New York: W. W. Norton & Co. 1993

Hatab, Lawrence. *Nietzsche and Eternal Recurrence: The Redemption of Time and Becoming.* Lanham: Rowman & Littlefield. 1985

Hawkins, Jeff. *On Intelligence.* New York: Times Books. 2004

Hawkins, Harriett. *Strange Attractors: Literature, Culture, and Chaos Theory.* New York: Prentice Hall/Harvester Wheatsheaf. 1995

Hayek F. A. *Individualism and Economic Order.* Chicago: University of Chicago Press. 1948

Hayles, N. Katherine, ed. *Chaos and Order: Complex Dynamics in Literature and Science.* Chicago and London: The University of Chicago Press. 1991

Hegel, Georg Wilhelm Friedrich. *The Philosophy of History.* J. Sibree, tr. New York: Dover Publications. 1956

Heidegger, Martin. *Introduction to Metaphysics.* New Haven & London: Yale University Press. 2000

———. *Poetry, Language, Thought.* New York: Harper & Row. 1971

Holland, John. H. *Hidden Order.* New York: Helix Books. 1995

Hubbard, Barbara Burke. *The World According to Wavelets.* Natick, MA: A. K. Peters. 1998

Huizinga, Johan. *homo ludens.* Boston: The Beacon Press. 1950

Hutcheson, Francis. *An Inquiry into the Original of Our Idea of Beauty and Virtue.* Indianapolis: Liberty Fund. 2004

Johnson, Steven. *Emergence.* New York: A Touchstone Book. 2001

Kahn, Charles H. *The Art and Thought of Heraclitus.* Cambridge: Cambridge University Press. 1979

Kainz, Howard P. *Paradox, Dialectic, and System.* University Park and London: The Pennsylvania State University Press. 1988

Kauffman, Stuart. *At Home in the Universe.* New York: Oxford University Press. 1995

———. *Investigations.* New York: Oxford University Press. 2000

———. *The Origins of Order.* New York: Oxford University Press. 1993

Kessler, M. A. and B. T. Werner. "Self-Organization of Sorted Patterned Ground" *Science* 17 January 2003

Klein, Étienne. *Chronos.* New York: Thunder's Mouth Press. 2005

Kurzweil, Ray. *The Singularity Is Near.* New York: Viking. 2005

Kurzweil, Ray. www.kurzweilai.net/articles/art0134.html?printable:1

Laszlo, Ervin. *The Systems View of the World.* New York: George Braziller. 1972

Laughlin, Robert B. *A Different Universe.* New York: Basic Books. 2005

Levin, Janna. *How the Universe Got Its Spots.* London: Phoenix. 2003

Lloyd, Seth. *Programming the Universe.* New York: Alfred Knopf. 2006

Lopreato, Joseph. *Human Nature and Biocultural Evolution.* Boston: Allen & Unwen. 1984

Lumsden, Charles J. and Edward O. Wilson. *Promethean Fire: Reflections on the Origin of Mind.* Cambridge: Harvard University Press. 1983

———. *Genes, Mind, and Culture.* Cambridge: Harvard University Press. 1981

Malin, Shimon. *Nature Loves to Hide.* Oxford: Oxford University Press. 2001

Mandelbrot, Benoit B. *The Fractal Geometry of Nature.* W. H. Freeman and Co. 1977

Mann, Daniel. "On Patterned Ground" *Science* 17 Jan 2003

McNamara, Kenneth J. *Shapes of Time.* Baltimore, MD: Johns Hopkins University Press. 1997

Mezard, Marc. "Passing Messages Between Disciplines" *Science*. 19 Sept. 2003

Michon, John A., et al., ed. *Guyau and the Idea of Time*. Amsterdam: North-Holland Publishing Company. 1998

Neville, Robert Cummings. *Eternity and Time's Flow*. Albany: SUNY Press. 1993

Newton, Eric. *The Meaning of Beauty*. Middlesex: Penguin Books. 1962

Nietzsche, Friedrich. *The Birth of Tragedy*. New York: Penguin Books. 1993

———. *The Gay Science*. New York: Random House. 1974

———. *Human, All Too Human*. Lincoln and London: University of Nebraska Press. 1986

———. *On the Genealogy of Morals and Ecce Homo*. New York: Random House, Inc. 1967

———. *Philosophy and Truth*. New Jersey: Humanities Press International, Inc. 1979/1990

———. *Philosophy in the Tragic Age of the Greeks*. Washington, D.C.: Regnery Publishing, Inc. 1962/1996

———. *Thus Spoke Zarathustra*. New York: Penguin Books. 1969

———. *Twilight of the Idols / The Anti-Christ*. New York: Penguin Classics. 1990

———. *The Will to Power*. New York: Vintage Books. 1968

Odenwald, Sten F. *Patterns in the Void*. New York: Westview Press. 2002

O'Donohue, John. *Beauty: The Invisible Embrace*. New York: HarperCollins. 2004

Orlock, Carol. *Inner Time*. New York: Birch Lane Press. 1993

Parker, Barry. *Chaos in the Cosmos*. New York: Plenum Press. 1996

Paulos, John Allen. *Innumeracy*. New York: Hill and Wang. 1988

Peitgen, Heinz-Otto, et al. *Chaos and Fractals: New Frontiers in Science*. New York: Springer-Verlag. 1992

Penrose, Roger. *The Road to Reality*. New York: Alfred A. Knopf. 2005

Pierce, John R. *An Introduction to Information Theory*. New York: Dover Publications. 1980

Bibliography

Price, Huw. *Time's Arrow and Archimedes' Point*. Oxford: Oxford University Press. 1996

Prigogine, Ilya. *The End of Certainty*. New York: The Free Press. 1997

———. *From Being to Becoming*. New York: W. H. Freeman and Co. 1980

Prigogine, Ilya & Isabelle Stengers. *Order Out of Chaos*. New York: Bantam Books. 1984

Pinker, Steven. *The Stuff of Thought*. New York: Viking. 2007

Quine, Willard Van Orman. *Elementary Logic*. Cambridge, MA: Harvard University Press. 1998

———. *From A Logical Point of View*. Cambridge, MA: Harvard University Press. 1953

———. *Philosophy of Logic*. Cambridge, MA: Harvard University Press. 1970

———. *Pursuit of Truth*. Cambridge, MA: Harvard University Press. 1992

Randall, Lisa. *Warped Passages*. New York: HarperCollins Publisher. 2005

Russell, Bertrand. *The Principles of Mathematics*. New York: W. W. Norton & Co. 1996

Sabelli, Héctor. *Bios*. New Jersey: World Scientific. 2005

Sandefur, James T. *Discrete Dynamical Systems*. Oxford: Oxford University Press. 1990

Santayana, George. *The Sense of Beauty*. New York: Dover Publications. 1955

Sartwell, Crispin. *Six Names of Beauty*. New York: Routledge. 2004

Satinover, Jeffrey. *The Quantum Brain*. New York: John Wiley & Sons. 2001

Scarry, Elaine. *On Beauty and Being Just* Princeton and Oxford: Princeton University Press. 1999

Schneider, Eric D. and Dorion Sagan. *Into the Cool*. Chicago: University of Chicago Press. 2005

Seife, Charles. *Decoding the Universe*. New York: Viking Press. 2006

———. "Doing the Wave in Many Ways," *Science* 15 Nov. 2002

Skutch, Alexander F. *Origins of Nature's Beauty*. Austin: University of Texas Press. 1992

Slobodkin, Lawrence B. *Simplicity and Complexity in Games of the Intellect*. Cambridge, MA: Harvard University Press. 1992

Smoot, George and Keay Davidson. *Wrinkles in Time*. New York: William Morrow & Co. 1993

Stamovlasis, Dimitrios. "The Nonlinear Dynamical Hypothesis in Science Education Problem Solving: A Catastrophe Theory Approach" *Nonlinear Dynamics, Psychology, and Life Sciences*. 10(1) 2006

Stewart, Ian. *Does God Play Dice?* Malden, MA: Blackwell Publishing. 1989

Susskind, Leonard. *The Cosmic Landscape*. New York: Little, Brown and Company. 2006

Taylor, Mark C. *The Moment of Complexity*. Chicago: The University of Chicago Press. 2001

Taylor, Richard P. "Order in Pollock's Chaos" *Scientific American*, Dec. 2002

Turner, Frederick. *Beauty*. Charlottesville and London: University Press of Virginia. 1991

———. *The Culture of Hope*. New York: The Free Press. 1995

———. *Natural Classicism*. Charlottesville: University Press of Virginia. 1992

Waldrop, M. Mitchell. *Complexity*. New York: Simon & Schuster. 1993

Waugh, Alexander. *Time*. London: Headline Book Publishing. 1999

Wechsler, Judith, ed. *On Aesthetics in Science*. Boston and Basel: Birkhäuser. 1988

Whitehead, Alfred North. *Process and Reality*. New York: The Free Press. 1985

Whitrow, G. J. *Time in History*. New York: Barnes & Noble Books. 1988

———. *What is Time?* Oxford: Oxford University Press. 1972

Weiner, Norbert. *Cybernetics: or Control and Communication in the Animal and the Machine*. Cambridge, MA: The MIT Press. 1948

Wilson, E. O. *Consilience: The Unity of Knowledge*. New York: Alfred A. Knopf. 1998

Wolfram, Stephen. *A New Kind of Science*. Winnipeg: Wolfram Media, Inc. 2002

Wright, Robert. *Nonzero: The Logic of Human Destiny*. New York: Vintage Books. 2001

Index

(by Section)

Abstraction, 8, 13, 19, 24, 29, 79, 80, 104
Action, 6, 28, 37, 46, 48, 58, 67, 82, 85, 93, 106, 113, 128, 145, 147, 166, 198, 212, 243, 245, 258, 289
Aeschylus, *see Tragedy*
Anarchy, 38, 67, 81, 296
Aristotle, 32, 34, 65, 126, 207, 210, 212, 219
Art(s) & Artists, 4, 5, 6, 7, 8, 33, 34, 35, 38, 40, 60, 62, 66, 67, 68, 70, 81, 96, 162, 194, 200, 210, 237, 264, 266
Atoms, 16, 35, 40, 57, 68, 70, 73, 75, 79, 84, 85, 86, 88, 95, 105, 106, 108, 112, 117, 128, 142, 143, 149, 151, 158, 173, 174, 175, 176, 188, 198, 200, 224, 228, 235, 246, 247, 248, 251, 252, 256, 271, 273, 276
Attraction (and Repulsion), 63, 158, 228, 247, 248, 249, 252
Attractors, 56, 179, 180, 247, 248, 251
  Point, 34, 62, 63
  Strange/Chaotic/Biotic, 36, 37, 62, 63, 101, 123, 196, 197, 200, 248, 249, 250, 257, 260, 267
Beauty, 24, 25, 67, 105, 126, 189, 211, 212, 213, 214, 223, 230, 233, 286, 300
Beck, Don and Christopher Cowan, 65, 68, 151, 154, 155, 195, 278, 282, 283, 284, 285, 286, 289, 291
Bertalanffy, Ludwig von, 293, 295, 296
Bifurcat(ion), 48, 63, 84, 88, 100, 179, 193, 196, 197, 247, 251, 252, 260, 266
Biology, 30, 33, 35, 36, 38, 41, 42, 60, 65, 66, 67, 68, 70, 84, 105, 119, 123, 128, 135, 148, 156, 157, 158, 160, 161, 176, 178, 252, 255, 266, 269, 276, 279, 280, 281, 284, 287, 298
Bios, 36, 52, 53, 56, 64, 68, 100, 133, 138, 171, 182, 196, 251, 284
Blake, William, 201, 205, 215
Bonner, John T., 71, 73
Brain, 4, 9, 10, 11, 21, 24, 50, 71, 72, 73, 86, 95, 106, 113, 125, 128, 130, 160, 173, 174, 178, 188, 194, 195, 196, 197, 200, 206, 237, 266, 272, 285
Burgin, Mark, 57, 58
Cancer, 206, 208, 210, 212
Casti, John, 33, 36, 41
Catastrophe Theory, 33, 36, 115
Causality, 68, 139, 222
Chaos, 36, 62, 63, 65, 68, 99, 159, 160, 171, 182, 199, 215, 229, 252, 265, 267, 272, 286, 289, 291
Campbell, Jeremy, 81, 83, 135, 140, 182, 243
Chemistry, 4, 10, 23, 30, 35, 42, 60, 64, 68, 70, 84, 88, 95, 99, 105, 106, 108, 116, 123, 128, 143, 148, 149, 151, 152, 153, 157, 158, 161, 173, 175, 176, 178, 199, 200, 206, 209, 224, 228, 235, 237, 246, 256, 257, 258, 267, 269, 270, 271, 272, 273, 274, 275, 276, 279, 280, 281, 284, 286, 287
Cherfas, Jeremy, *see Gribbin, John*
Chimpanzee, 13, 14, 38, 154, 266, 269, 278
Communication, 4, 38, 49, 72, 84, 105, 157, 182, 186, 188, 206, 223, 254, 255, 287, 288
Community, 72, 197, 211, 214
Complexity, 11, 13, 15, 29, 30, 33, 35,

40, 48, 52, 53, 54, 55, 60, 63, 64, 65, 66, 68, 69, 70, 71, 72, 73, 77, 81, 84, 88, 93, 94, 95, 96, 97, 98, 99, 100, 101, 102, 103, 105, 107, 109, 116, 119, 129, 134, 136, 138, 143, 145, 146, 148, 149, 151, 152, 153, 154, 155, 156, 157, 158, 160, 161, 164, 165, 166, 167, 168, 173, 174, 175, 176, 177, 182, 191, 192, 195, 198, 199, 200, 206, 211, 214, 216, 223, 226, 227, 235, 236, 237, 241, 243, 244, 246, 247, 248, 249, 250, 251, 252, 253, 255, 256, 257, 258, 259, 262, 265, 266, 267, 269, 270, 271, 272, 273, 274, 275, 276, 277, 278, 280, 281, 282, 283, 284, 285, 286, 287, 288, 289, 290, 291, 292, 295, 299
Concepts and Conceptual Categories, 4, 8, 10, 11, 13, 14, 15, 17, 18, 19, 20, 21, 22, 24, 26, 27, 31, 62, 66, 72, 79, 80, 104, 201, 233, 285
Cowan, Christopher, *see Beck, Don*
Creat(ivity)/Creation, 5, 10, 11, 14, 18, 20, 22, 35, 36, 37, 40, 41, 42, 45, 48, 52, 56, 57, 62, 63, 64, 66, 67, 68, 70, 71, 72, 73, 75, 76, 77, 79, 80, 81, 84, 86, 93, 95, 96, 100, 104, 105, 107, 108, 109, 116, 120, 125, 130, 132, 133, 140, 149, 157, 158, 160, 171, 173, 174, 182, 185, 192, 193, 194, 195, 195, 197, 198, 200, 204, 208, 210, 212, 1213, 215, 216, 224, 226, 227, 228, 230, 239, 244, 245, 247, 248, 249, 251, 252, 254, 255, 256, 257, 258, 259, 260, 262, 263, 266, 267, 270, 271, 280, 281, 283, 286, 288, 289, 290, 291
Culture, 4, 6, 33, 34, 38, 40, 60, 65, 68, 71, 125, 144, 154, 158, 160, 194, 195, 199, 200, 205, 210, 212, 237, 266, 267, 278, 279, 280, 283, 284, 291, 295
Cybernetics, *see Feedback*
Darwin, Charles, 41, 65, 105, 274
Derrida, Jacques, 32
Descartes, Rene, 7, 203
Determinism, 17, 40, 62, 63, 64, 88, 92, 99, 114, 134, 154, 157, 161, 273, 274, 278, 281, 284
Developmental Biology/Genetics, 135, 260, 261

Diaphysics, 37, 39, 40, 43, 99, 115, 119, 145, 299, 300
Digital-Analog, 37, 70, 77, 78, 85, 86, 107, 22,3, 225, 281
Disorder, *see Order*
Dissipative Structure, 53, 64, 65, 73, 74, 76, 86, 88, 100, 106
DNA/Gene(tics)/Molecular Biology, 50, 66, 73, 125, 135, 149, 160, 194, 195, 196, 212, 215, 236, 254, 266
Ecology, 155, 170, 181
Economics, 4, 35, 40, 57, 60, 66, 144, 145, 158, 170, 172, 178, 228, 266, 291, 295, 297, 298
Education, 290, 298, 299
Einstein, Albert, 26, 40, 94, 99, 273
Emergence, 4, 16, 35, 36, 37, 44, 57, 59, 60, 64, 65, 66, 68, 69, 70, 76, 77, 84, 86, 88, 94, 95, 96, 97, 98, 99, 100, 105, 106, 107, 120, 128, 129, 133, 135, 138, 139, 144, 145, 146, 147, 148, 150, 154, 155, 156, 157, 160, 161, 164, 173, 176, 178, 179, 185, 190, 200, 216, 223, 238, 239, 243, 244, 258, 266, 267, 269, 270, 271, 273, 274, 275, 276, 277, 278, 279, 280, 281, 282, 283, 284, 285, 287, 288, 289, 290, 291, 292
Energy, 21, 45, 46, 49, 52, 53, 57, 58, 60, 64, 71, 74, 78, 79, 95, 97, 99, 123, 136, 150, 166, 173, 175, 183, 185, 191, 192, 200, 226, 239, 245, 247, 251, 252, 254, 255, 259, 270, 271, 273, 276, 279, 280, 281
Entropy, 45, 47, 52, 53, 54, 55, 64, 71, 72, 74, 86, 97, 100, 179, 191, 239, 244, 247, 252, 253, 254, 255, 293
Environment, 50, 68, 73, 84, 109, 125, 133, 159, 160, 161, 195, 198, 208, 259, 280, 283, 290
Epistemology, 7, 52, 53, 124, 126
Eukaryote(s), 84, 149, 153, 195, 236, 259, 260, 264, 267, 276
Evolution
    Biological/Natural Selection, 38, 52, 65, 88, 95, 103, 109, 116, 125, 133, 134, 135, 155, 159, 172, 194, 195, 259, 260, 262, 265, 267, 283
    Cultural, 38, 65, 68, 196
    Human, 30, 38, 39, 42, 65, 68, 95, 109, 117, 119, 136, 155,

Index

157, 174, 266, 267, 272, 283, 285, 298
Physical, 36, 52, 56, 64, 65, 66, 68, 95, 96, 99, 100, 147, 166, 180, 182, 223, 227, 246, 252, 258, 267, 273, 275
Existentialism, 32, 62
Extinction, 84, 93, 159, 161, 172, 195, 198, 266
Far-From-Equilibrium, 134, 171, 172, 193, 194, 195, 200, 252, 255, 256, 257, 258, 259, 266, 267, 268
Feedback, 56, 68, 70, 125, 157, 177, 182, 223, 279, 280
Fibonacci series/spiral, *see Golden Mean*
Fitness Landscapes, 156, 159, 160, 161, 252, 257, 259, 265, 266, 267, 278
Fold(ing), 62, 93, 98, 149, 166, 173, 174, 175, 176, 184, 186, 190, 260, 271, 273, 274, 288, *see also Waves/Solitons*
Fractal, 14, 59, 60, 61, 62, 67, 98, 99, 101, 107, 108, 109, 114, 115, 141, 143, 146, 174, 182, 229, 233, 244, 252, 262, 271, 272, 275, 286, 292
Fraser, J.T., 65, 68, 87, 88, 94, 95, 105, 106, 150, 155, 190, 226, 243, 273, 277, 283, 285
Freedom 38, 62, 66, 71, 80, 81, 128, 129, 209, 210, 211, 239, 255, 266, 268, 274 285, 288, 291
Frege, Gottlob, 12, 14, 15, 16, 17, 20, 23, 27
and Jevons, W. S., 25
Fuchs, Christian, 64, 223
Game Theory 8, 65, 66, 71
Gefter, Amanda, 235
Gene(tics), *see DNA*
Golden Mean/Fibonacci series/spiral, 62, 131, 230, 232, 257
Gould, Jay, 65
Government/Politics, 33, 35, 38, 40, 60, 66, 124, 172, 199, 204, 290, 291, 298
Graves, Clare, *see Beck, Don*
Gravity, 40, 62, 63, 97, 174, 186, 187, 190, 235, 247, 251
Gribbin, John and Jeremy Cherfas, 65
Growth, 53, 64, 78, 82, 112, 145, 171, 172, 176, 230, 293
Hawkins, Harriett, 62

Hawkins, Jeff, 86, 87, 109, 120
Health, 43, 109, 126, 197, 201, 202, 206, 208, 209, 210, 211, 212, 213, 214, 290
Hegel, G. W. F., 203
Heidegger, Martin, 32, 41, 62, 71, 124, 132
Heraclitus, 26, 34, 37, 45, 64, 71, 72, 212, 214
Hierarchy, 11, 13, 33, 54, 64, 65, 70, 72, 73, 84, 99, 105, 107, 108, 109, 144, 146, 154, 155, 156, 157, 206, 209, 211, 223, 267, 278, 279, 283, 284, 286, 295
History and Historicism, 15, 90, 102, 124, 204, 223, 281, 284, 285, 296
Holy, 3, 201, 203, 204, 205, 206, 209, 211, 212, 213, 214
Human Nature, 34, 42, 291
Humanities, 3, 4, 41, 164
Hutcheson, Francis, 25
Infinity, 47, 61, 62, 72, 141, 142, 143, 174, 201, 205, 215, 221, 229, 231, 232, 233, 234, 252, 271
Information and Information Theory, 4, 10, 12, 13, 19, 20, 28, 36, 37, 38, 44, 45, 46, 47, 48, 49, 50, 52, 53, 54, 56, 57, 58, 60, 62, 68, 71, 72, 73, 74, 75, 76, 77, 78, 83, 86, 95, 97, 99, 105, 132, 133, 135, 140, 158, 167, 168, 180, 182, 183, 185, 186, 187, 188, 191, 192, 210, 217, 223, 235, 237, 250, 255, 257, 284, 287, 299
Intelligence, 38, 70, 86, 87, 92, 100, 105, 108, 109, 120, 157, 167, 173, 174, 246, 270, 273, 274, 277, 282
Interaction, 4, 30, 48, 64, 68, 69, 76, 77, 83, 84, 85, 86, 95, 97, 98, 104, 107, 128, 140, 155, 173, 182, 185, 186, 187, 188, 191, 192, 223, 223, 235, 236, 237, 248, 252, 256, 270, 271, 272, 283
Interdisciplinary Thinking, 135, 284, 286
Jevons, W. S., *see Frege, Gottlob*
Kant, Immanuel, 7, 203
Kauffman, Stuart, 159, 244, 249, 250, 254, 255, 258, 266, 268, 270, 272
Knowledge, 1, 6, 7, 9, 15, 31, 42, 52, 53, 58, 72, 82, 92, 95, 96, 98, 101, 102, 103, 108, 117, 118, 119, 121,

126, 127, 130, 191, 192, 209, 210, 215, 226, 240, 241, 255, 279, 284, 286, 288, 289, 291
Kurzweil, Ray, 86
Language, 6, 11, 14, 15, 17, 20, 21, 38, 40, 41, 42, 60, 66, 84, 95, 99, 132, 157, 158, 194, 195, 200, 266, 279, 287, 288
Laplace, Pierre-Simon, 62
Laughlin, Robert B., 144, 145, 146, 246
Laws, 27, 34, 35, 37, 39, 45, 66, 97, 104, 119, 144, 145, 147, 191, 199, 214, 239, 244, 245, 246, 247, 251, 252, 267, 286, *see also Rules*
Linear, *see Nonlinear*
Literature, 33, 35, 38, 40, 60., 66, 67, 210, 237, 266
Lloyd, Seth, 83, 97, 185, 191, 192, 225, 243, 250
Logos, 4, 37, 38, 39, 44, 45, 59, 71, 72, 170, 214
Lorenz, Ralph, 64
Love (and Strife), 29, 158, 209, 210, 211, 227, 228
Lysenko, Trofim, 5, 41
Malin, Shimon, 76, 77, 79, 82, 102, 103
Marxism (and Communism), 41, 198, 203, 204, 205, 278
Mathematics, 8, 9, 10,20, 21, 24, 25, 29, 30, 33, 61, 64, 84, 102, 103, 114, 144, 159, 189, 229, 231, 233, 238, 247, 281
Matter, 45, 48, 49, 52, 58, 61, 62, 70, 72, 75, 97, 144, 166, 173, 175, 185, 223, 246
Metahuman, 200, 269, 284, 285, 286, 287, 288, 289, 290, 291
Metaphor, 6, 35, 189, 288
Metaphysics, 32, 33, 34, 35, 36, 37, 41, 43
Mézard, Marc, 57
Mill, J.S., 14, 18, 20, 23
Mind, 4, 15, 17, 22, 33, 35, 77, 79, 106, 114, 125, 128, 160, 194, 195, 196, 197, 200, 210, 211, 240, 241, 246, 266, 287
Molecular Biology, *see DNA*
Morowitz, Harold J., 116
Nature, 4, 10, 21, 24, 27, 28, 34, 56, 60, 62, 64, 66, 100, 103, 104, 107, 114, 119, 124, 125, 136, 145, 146, 213, 214, 246, 252, 281, 284, 286, 289, 291, 292
Nazis, 41
Neoteny, 148, 149, 152, 153, 155, 156, 195, 200, 258, 262, 266, 267, 283
Newton, Sir Isaac, 53, 62, 94, 95, 273
Nietzsche, 7, 8, 9, 26, 34, 51, 58, 62, 66, 102, 124, 203
Nomos, 4, 5, 33, 34, 37, 38, 39, 41, 42, 43, 59, 72, 122, 147, 170, 214
Nonlinear Dynamics (vs. Linear), 88, 135, 138, 157, 171, 172, 178, 223, 235, 236, 237, 248, 280, 281, 284, 285
Nothing, 7, 25, 37, 92, 102, 123, 142, 173, 176, 185, 245
Nouns (the problem of), 12, 89, 106, 111, 184
Number, 12, 13, 14, 15, 17, 18, 19, 20, 21, 22, 24, 25, 28, 33, 230, 231, 232
Ontology, 12, 45, 103, 185
Order (and Disorder), 16, 28, 29, 38, 40, 52, 53, 62, 64, 65, 68, 71, 82, 86, 88, 90, 97, 100, 202, 205, 106, 119, 129, 159, 160, 161, 170, 171, 172, 175, 182, 199, 214, 215, 221, 229, 236, 246, 249, 262, 26, 272, 286, 288, 289, 291, 293
Organism, 36, 39, 50, 53, 60, 73, 84, 96, 99, 106, 108, 125, 126, 133, 134, 135, 136, 138, 143, 149, 153, 159, 161, 171, 173, 179, 188, 197, 199, 206, 209, 212, 214, 236, 237, 255, 258, 259, 260, 262, 263, 271, 274, 280, 285
Paradox, 51, 70, 85, 105, 107, 110, 116, 216, 217, 218, 219, 220, 221, 222, 223, 224, 227, 228, 229, 230, 231, 232, 233, 234, 235, 236, 238, 243, 276, 283, 286, 289
Pattern(s), 141, 5, 20, 22, 36, 46, 52, 56, 62, 67, 68, 72, 77, 86, 87, 107, 108, 109, 113, 155, 173, 191, 215, 249, 260, 263, 283
Paulos, John Allen, 79
Perspectiv(ism), 47, 54, 144, 154, 192, 200, 278, 286, 289
Philosophy, 4, 9, 18, 21, 25, 26, 32, 33, 34, 35, 37, 40, 70, 72, 76, 87, 102, 103, 162, 176, 203, 204, 210, 298
Physics

Index

Mechanistic, 35, 40, 53, 84, 88, 94, 95, 103, 114, 152
Relativity, 40, 50, 53, 94, 99, 273
String Theory, 106, 114, 124, 185, 235, 236, 237
Quantum, 4, 24, 28, 40, 45, 47, 49, 50, 60, 66, 68, 70, 75, 76, 77, 78, 83, 84, 85, 86, 95, 97, 99, 100, 105, 107, 108, 116, 123, 128, 139, 149, 157, 185, 190, 222, 224, 225, 226, 235, 237, 250, 252, 255, 256, 258, 267, 269, 270, 271, 273, 274, 275, 276, 279, 280, 284, 285, 287, 292
Physis, 4, 5, 7, 26, 33, 34, 35, 37, 38, 39, 40, 41, 42, 59, 64, 71, 72, 96, 119, 122, 124, 125, 128, 130, 147, 170, 214
Pierce, John, 24, 254, 255
Pinker, Steven, 15
Poetry, 6, 86, 189, 201, 215, 261, 266, 288
Politics, *see Government*
Pollock, Jackson, *see Taylor, Richard*
Pöppel, Ernst, *see Turner, Frederick*
Postmodernism, 2, 32, 62
Potential, 47, 77, 78, 79, 82, 83, 93, 100, 103, 185, 196, 200, 288
Power Laws, 46, 86, 115, 172, 177, 182, 203, 204, 205, 206, 284
Prigogine, Ilya, 65, 256
Probability, 79, 92, 95, 97, 99, 103, 118, 134, 185, 242, 273, 274, 281
Prokaryote(s), 84, 153, 236, 259, 269, 280
Psychology, 21. 298
Quine, W. V. O., 12, 13
Reason(able), 1, 2, 3, 7, 17, 26, 51, 66, 101, 214
Reductionism, 96, 144, 240, 284
Redundancy, 72, 81, 264, 295
Relativism (Ethics), 40, 118, 124, 125, 154, 278, 243
Religion, 3, 60, 195, 196, 205, 212, 298
Repulsion, *see Attraction*
Rhythm(s), 66, 67, 86, 215, 223
Rules, 19, 30, 35, 36, 37, 40, 46, 47, 66, 67, 71, 81, 100, 101, 115, 116, 125, 127, 140, 144, 145, 182, 206, 209, 210, 223, 224, 239, 243, 246, 249, *see also Laws*

Sabelli, Hector, 28, 29, 46, 52, 53, 56, 68
Satinover, Jeffrey, 47, 149
Schroeder, Manfred, 114
Seife, Charles, 49, 50, 226
Self-Organiz(ation), 46, 64, 86, 115, 133, 140, 177, 181, 182, 215, 223, 257, 268, 292, 295
Self-Referentiality, 231, 235, 243
Self-Similar(ity), 61, 62, 99, 107, 115, 146, 150, 158, 182, 211, 223, 233, 235, 237, 244, 271, 272, 275
Shakespeare, William, *see Tragedy*
Skepticism, 1, 3, 126
Smith, John Maynard, 66
Social Construction (of Reality), 5
Society/Sociology, 3, 4, 15, 30, 33, 65, 66, 70, 84, 105, 123, 154, 155, 158, 178, 200, 266, 271, 276, 277, 278, 279, 283, 290, 291, 295, 296, 298
Solitons, *see Waves*
Sophocles, *see Tragedy*
Spacetime, 67, 77, 78, 82, 93, 94, 97, 98, 102, 142, 166, 173, 175, 176, 184, 185, 186, 187, 190, 238, 270, 271, 273, 274, 276, 288, 292
Strauss, Leo, 119
Strife, *see Love*
Symmetry and Symmetry-Breaking, 52, 144, 193, 200, 213, 230, 245, 246, 252, 256, 257, 258, 2359, 266, 267, 268
Systems, 15, 33, 40, 47, 49, 57, 58, 62, 63, 65, 74, 77, 81, 83, 84, 85 ,87, 95, 99, 103, 104, 108, 109, 114, 128, 132, 135, 138, 143, 146, 153, 157, 171, 173, 174, 179, 180, 181, 182, 187, 188, 191, 193, 195, 196, 197, 198, 199, 200, 206, 223, 227, 235, 236, 237, 243, 247, 248, 250, 254, 255, 257, 258, 262, 274, 276, 286, 294, 297, 299
Closed, 45, 53, 64, 253
Open, 37, 53, 60, 64, 66, 70, 71, 72, 73, 76, 86, 88, 96, 97, 107, 115, 130, 134, 140, 147, 149, 152, 165, 171, 172, 198, 220, 223, 238, 243, 249, 251, 252, 256, 267, 268, 270, 273, 291, 293, 295
Taylor, Richard, 62
Technology, 38, 41, 42, 70, 96, 125, 130, 160, 195, 200, 237, 266, 267

Thermodynamics, Law of, 45, 50
  First, 45, 51, 191
  Second, 45, 41, 191, 247, 251, 253, *see also Entropy*
Time, 13, 27, 40, 46, 58, 62, 64, 67, 68, 69, 78, 89, 101, 119, 142, 143, 146, 171, 172, 184, 185, 192, 223, 226, 234, 244, 255, 258, 261, 268, 292, 296, 297, *see also Spacetime*
  umwelt theory of, 65, 68, 73, 93, 94, 95, 96, 99, 102, 155, 190, 273, 281, 283, 284, 285
  Present/Past/Future, 5, 20, 38, 42, 43, 62, 90, 91, 92, 93, 95, 101, 103, 195, 255, 257, 273, 274, 281, 282 288, 292, 296
Tragedy, 5, 7, 41, 43, 59, 60, 65, 93, 120, 124, 214, 280
Transcendentalism, 8, 9, 20, 21, 24, 25, 26, 31, 77, 79, 80, 107
Truth, 3, 101, 102, 118, 125, 154, 203, 278
  and Lies, 16, 26, 215, 297
  as Facts, 9, 24, 76, 118, 125, 133
Turner, Frederick, 54, 93, 288
  and Ernst Pöppel, 190
Universe, 13, 14, 23, 24, 37, 40, 45, 46, 47, 48, 52, 53, 54, 56, 58, 61, 62, 69, 66, 86, 94, 95, 96, 97, 98, 101, 104, 107, 109, 117, 119, 131, 132, 139, 142, 143, 144, 146, 149, 155, 157, 158, 167, 173, 177, 179, 188, 192, 193, 198, 200, 201, 216, 223, 227, 233, 234, 235, 236, 237, 238, 244, 245, 246, 247, 249, 250, 251, 252, 253, 254, 255, 256, 258, 262, 268, 270, 273, 274, 275, 283, 286, 288, 289, 292, 293, 300
Wald, George, 117
Waves/Solitons, 40, 56, 60, 70, 76, 77, 78, 83, 85, 86, 95, 97, 102, 103, 105, 106, 108, 110, 128, 149, 151, 168, 169, 173, 175, 176, 184, 185, 186, 188, 190, 191, 192, 222, 225, 235, 252, 256, 258, 271, 273, 276, 280, 286, 287, *see also Fold(ing)*
Wisdom, 37, 126, 204, 210, 241, 286
Wittgenstein, Ludwig, 6
Wolfram, Stephen, 62, 255
Wright, Robert,

About the Author

Troy Camplin is an interdisciplinary scholar, poet and short story writer, president of The Emerson Institute for Freedom and Culture <http://www.emersoninstitute.org>, and Reviews Editor for *Philosophical Practice*. He maintains a blog, Interdisciplinary World <http://www.zatavu.blogspot.com/>, where he continues developing some of the ideas of this book. *Diaphysics* is his first book.